AQA
GCSE (9–1)

ENGINEERING

Paul Anderson
David Hills-Taylor

Contributor:
Mark Griffiths

Approval message from AQA

This textbook has been approved by AQA for use with our qualification. This means that we have checked that it broadly covers the specification and we are satisfied with the overall quality. Full details of our approval process can be found on our website.

We approve textbooks because we know how important it is for teachers and students to have the right resources to support their teaching and learning. However, the publisher is ultimately responsible for the editorial control and quality of this book.

Please note that when teaching the *AQA GCSE Engineering* course, you must refer to AQA's specification as your definitive source of information. While this book has been written to match the specification, it cannot provide complete coverage of every aspect of the course.

A wide range of other useful resources can be found on the relevant subject pages of our website: www.aqa.org.uk.

HODDER
EDUCATION
AN HACHETTE UK COMPANY

Although every effort has been made to ensure that website addresses are correct at time of going to press, Hodder Education cannot be held responsible for the content of any website mentioned in this book. It is sometimes possible to find a relocated web page by typing in the address of the home page for a website in the URL window of your browser.

Hachette UK's policy is to use papers that are natural, renewable and recyclable products and made from wood grown in well-managed forests and other controlled sources. The logging and manufacturing processes are expected to conform to the environmental regulations of the country of origin.

Orders: please contact Hachette UK Distribution, Hely Hutchinson Centre, Milton Road, Didcot, Oxfordshire, OX11 7HH. Telephone: +44 (0)1235 827827. Email education@hachette.co.uk Lines are open from 9 a.m. to 5 p.m., Monday to Friday. You can also order through our website: www.hoddereducation.co.uk

ISBN: 9781510425712

© Paul Anderson and David Hills-Taylor 2018

First published in 2018 by
Hodder Education,
An Hachette UK Company
Carmelite House
50 Victoria Embankment
London EC4Y 0DZ

www.hoddereducation.co.uk

Impression number   10 9 8 7 6 5

Year   2022

Cover photo © Oliver Tindall / Alamy Stock Photo

Illustrations by Integra Software Serv. Ltd

Typeset in India

Printed and bound by CPI Group (UK) Ltd, Croydon, CR0 4YY

A catalogue record for this title is available from the British Library.

CONTENTS

Answers to Check your knowledge and understanding questions and Practice Questions available at www.hoddereducation.co.uk/AQAGCSEEngineering

Introduction

Developments in engineering have had a massive impact on every aspect of our daily lives: the buildings we live and work in, the cars we drive and the computers, smart phones and tablets we use have all been developed as a result of engineering. Engineering ideas are constantly evolving and new engineering innovations developed, making this a dynamic and exciting area to study and in which to work.

Engineering activities takes place in many different industries – manufacturing, design, aerospace, fabrication, motor vehicle, electronics, mechanics, food manufacture, clothing manufacture and communications all make use of engineering skills.

Throughout your GCSE Engineering course you will gain an insight into some of the aspects that make engineering such an interesting subject. You will be given the opportunity to develop a knowledge and understanding of wide range engineering ideas, techniques and skills, including:

- engineering materials, their properties and characteristics
- manufacturing processes and techniques, and how they are carried out
- the use and role of the mechanical, electrical, electronic, structural and pneumatic systems within engineering settings
- a range of testing and investigation methods that engineer's use
- the impact that modern technologies and the engineering industry have on production, society and the environment
- how to apply a range of engineering skills to help solve practical engineering problems.

Photo credits

How to use this book

This book has been written to support you through your AQA GCSE Engineering course. It is divided into seven sections.

Sections 1–6 cover the six sections of subject content in the AQA specification. In these sections you will find a range of different features that have been designed to help

you learn and improve your knowledge and understanding of engineering.

Section 7 looks at the non-exam assessment (NEA) and written examination in more detail. It provides advice on how to set about preparing and completing your project, and how to revise and prepare for the written exam.

What will I learn?
Clear learning objectives for each topic explain what you need to know and understand.

Activity
Short activities are included to help you to understand what you have read. Your teacher may ask you to complete these.

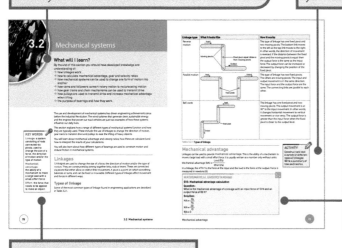

Mathematical understanding
These boxes include worked examples that guide you through how to apply engineering equations and mathematical skills and knowledge.

Key words
All important terms are defined.

Practice questions
These questions appear at the end of each section and are designed to help you to revise.

Stretch and challenge
These activities will help you to develop your understanding further. They may ask you to complete further research, or to consider more complex or challenging topics.

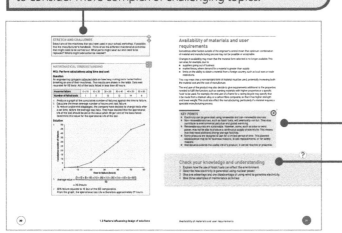

Key points
Short summaries of key points are included at the end of each topic to help you to remember what you have learnt.

Check your knowledge and understanding
These short questions test your knowledge and understanding of each topic.

1 Engineering materials

Successful engineered products do not just magically come into existence from nowhere. Normally, the needs that the product must meet are identified and researched; then considerable thought goes into choosing a material (or materials) that can meet these needs.

There are different categories of materials, such as metals, polymers and composites; and within each individual category, the different materials have unique combinations of properties, which make them the best for different applications.

This section includes the following topics:

1.1 Materials and their properties

1.2 Material costs and supply

1.3 Factors influencing design of solutions

At the end of this section you will find practice questions relating to engineering materials.

1.1 Materials and their properties

What will I learn?

By the end of this chapter you should have developed knowledge and understanding of:

→ the meaning of the different properties of materials
→ the different types of metal and how the properties of metals can be changed
→ thermoplastic and thermosetting polymers and how these are affected by heat
→ composite materials and how they are reinforced
→ other materials used in engineering, such as structural timber and ceramics.

There is a wide, and increasing, range of materials available to the engineer. These materials are most commonly referred to as types that depend upon how they are obtained. For example, metals are obtained by heating ores quarried from the ground; polymers are made by chemical processes; timber comes from trees. Each type typically has some similar characteristics, although each material within a type normally has slight differences in its combination of properties.

Material properties

Choosing the best material for an application means matching the properties needed by the application to the properties of the material. To be able to select the correct material, it is therefore essential to understand what the different properties mean. These include:

- strength
- ductility
- malleability
- hardness
- toughness and brittleness
- stiffness.

Strength

Strength is the ability of a material to withstand a force that is applied to it. There are different types of strength, depending upon the type of force applied to the material:

- Tensile strength is the ability to resist a pulling force.
- Compressive strength is the ability to resist a squeezing force.
- Torsional strength is the ability to resist a twisting force.

Some materials have good strength of all the different types, whereas others have good strength in one type, but are weaker in another. Ceramics, for example, typically have good compressive strength, but weaker tensile and torsional strength.

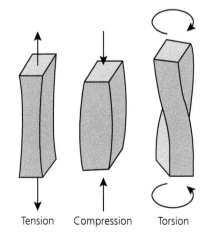

Tension   Compression   Torsion

Figure 1.1.1 Types of force

When considering the tensile strength of metals and polymers, often two different values are given. The **yield strength** is the amount of stress needed to start permanently deforming the material. Below the yield strength, the material stretches, but returns to its original size when the force is removed. Above this value the change in size (and shape) stays when the force is removed. The **ultimate tensile strength** is the stress at which the material eventually fails.

The strength of a product (rather than a material) depends upon both the type of the material it is made from and the area of the product over which the load is applied. This means that an applied force that has no visible effect on a large product may be enough to cause a small product to break. This is because with a smaller area, the stress that the force creates within the product may exceed what the material can stand.

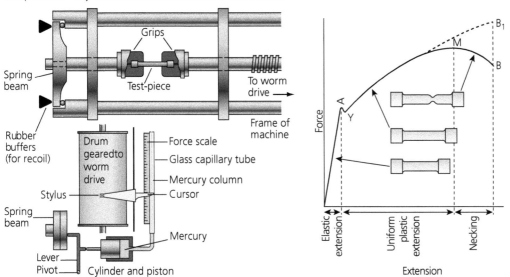

Figure 1.1.2 Testing the strengths of materials: material samples are placed in a vice and a load is applied

---

+ −
× =

MATHEMATICAL UNDERSTANDING

E7: Stress
(See section 4.1 for information on the conversion of load/extension to stress/strain and the formulae used.)

Question
A tensile test was carried out on a piece of metal. The test piece had a square section, with each side 10 mm. The force applied when the material started to yield was 24 500 newtons. Calculate the yield stress of the metal.

Solution

$$\text{Stress} = \frac{\text{force}}{\text{cross-sectional area}}$$

$$\sigma = \frac{F}{A}$$

Given force F = 24 500 N and cross-sectional area A = 10 x 10 = 100mm^2

$$\sigma = \frac{F}{A}$$

$$= \frac{24\,500}{100}$$

$$= 245 \text{ N mm}^{-2}$$

Ductility

Ductility is the amount that a material can be deformed. For example, most thermoplastic polymers are ductile. Nylon can be stretched by applying a tensile (pulling) force. On the other hand, ceramics are not ductile – they would resist the pulling force, but then they would fracture. One measure of ductility is the length that a piece of material extends when a load is applied, relative to its original length. As this is a ratio, this value does not have units.

MATHEMATICAL UNDERSTANDING

E8: Strain

(See section 4.1 for information on the conversion of load/extension to stress/strain and the formulae used.)

Question

A metal bar is being used as part of the lifting gear in a crane. When there is no load, the bar is 2 m long. When the crane lifts the maximum load, the bar extends to a length of 2.03 m.

Calculate the strain in the bar at the maximum load.

Solution

$$\text{Strain} = \frac{\text{change in length}}{\text{original length}}$$

$$\varepsilon = \frac{\Delta l}{l}$$

$$\varepsilon = \frac{0.03}{2} = 0.015 \text{ (or } 1.5 \times 10^{-2})$$

Figure 1.1.3
Hardness tester

Malleability

Malleability is the ability of a material to be deformed without rupturing. This means that the shape of the material can be changed without the material breaking. For example, modelling clay is very malleable; its shape can easily be changed by squeezing it with a hand. However, when it has dried out it is not malleable; attempting to reshape dried modelling clay results in it cracking and breaking.

Hardness

Hardness is the ability of a material to resist wear and abrasion. The harder a material, the more difficult it is to make a mark on its surface. It is also much more difficult to use saws or machines to cut the material. For example, if someone tries to scratch a piece of hard stainless steel, there may be at most only a tiny mark. If the same effort to scratch was applied to a piece of rubber, it would leave a much bigger indentation (mark).

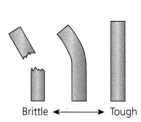

Brittle ← → Tough

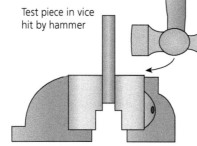

Test piece in vice hit by hammer

WARNING: Safety spectacles must be worn. Some materials may shatter.

Toughness/brittleness

Toughness is the ability of a material to withstand an impact without breaking. When it is hit with a hammer, a tough material might bend or be dented, but it will not crack. The opposite of toughness is **brittleness**. A brittle material will shatter in response to an impact. For example, glass is a brittle material. If a bowl made from glass is dropped it might smash. A similar bowl made from a tough metal might be dented when dropped.

Figure 1.1.4 Toughness testing : material samples are placed in a vice and hit with the same force using a hammer. Tough materials will not crack; brittle material will shatter.

Stiffness

Stiffness is the ability of a material to resist bending. In part, stiffness is related to the strength of the material – the stronger the material and the more it resists deforming, the stiffer it is. This is shown by the **Young's Modulus** of the material. However, stiffness is also strongly affected by the shape of the material. For example, an I-beam is stiffer than a round bar of the same volume.

MATHEMATICAL UNDERSTANDING

E9: Young's Modulus

Question

A tensile test was carried out on a ceramic test piece. The test piece was cylindrical in shape with a radius of 29.3 mm.

At the point when the applied force was 270 kN, the strain in the test piece was calculated to be 2.5×10^{-4}.

Calculate the Young's Modulus of the material.

Solution

Given

$$\text{Young's Modulus, } E = \frac{\text{stress, } \sigma}{\text{strain, } \varepsilon}$$

and

$$\text{Stress, } \sigma = \frac{\text{force, } F}{\text{cross-sectional area, } A}$$

$$\text{Cross-sectional area} = \pi r^2 = 2697 \text{ mm}^2$$

$$\text{Stress, } \sigma = \frac{270}{2697} \approx 0.1 \text{kN mm}^{-2}$$

$$\text{Young's Modulus, } E = \frac{0.1}{2.5 \times 10^{-4}} = 400 \text{kN mm}^{-2}$$

ACTIVITY

Identify five engineered products in your classroom or workshop. For each product, identify the properties that are important for the product to be able to carry out the task it was designed to do.

Metals and alloys

Metals are made from metal **ores**. The ores are rocks or minerals dug from quarries or mines then refined and processed, to turn the metal into usable forms.

Most metals are not used as pure chemical elements. They are typically mixed with other metals to improve their properties. A mixture of two or more metals is called an **alloy**. If should be noted that there are also a few non-metallic elements that can be added in small amounts to specific metals to form an alloy as well – for example, carbon in iron, or silicon in aluminium.

There are two main types of metals:
- **Ferrous** metals contain iron as their largest alloying element.
- **Non-ferrous** metals do not contain iron.

Figure 1.1.5 Pouring of liquid metal in a factory

Ferrous metals and alloys

Pure iron is too soft for use in most engineered products. Ferrous metals typically contain a small percentage of carbon, which makes them into an alloy called carbon steel. The amount of carbon has a significant effect on the properties of the alloy. In general, the more carbon, the harder and stronger the ferrous metal. Carbon steels are the most widely used metals in the world.

Compared to non-ferrous metals, carbon steels typically cost less. However, they are prone to corrosion when they are exposed to water, which causes them to rust. Other elements can be added to a ferrous alloy to reduce corrosion, although this increases the cost of the metal. Alternatively, a surface finish can be used, such as a protective coating.

Table 1.1.1 **Properties of ferrous alloys**

Ferrous alloy	Alloying elements include:	Properties	Typical uses include:
Cast iron	Typically 3–3.5% carbon	Good compressive strength. Hard, so can be difficult to machine, but suitable for casting. Brittle compared to other ferrous metals. Poor corrosion resistance, so rusts easily. Relatively low cost.	Anvils, engineering vices, engine blocks, machine tool beds
Low-carbon steel	Less than 0.3% carbon	Lower strength than other steels, but still stronger than most non-ferrous materials. Tough and relatively low cost. Cannot be hardened.	Nails and screws, car bodies, steel sheet
High-carbon steel	0.8–1.4% carbon	Strong and hard, but not as tough as lower carbon steel. Difficult to form. Can be hardened.	Tools, such as saw blades, hammers, chisels
Stainless steel	At least 11.5% chromium	Strong and hard. Difficult to machine. Good corrosion resistance: does not rust. Relatively expensive.	Knives and forks, medical equipment, sinks

KEY WORDS

Metal: a type of material typically made by processing an ore that has been mined or quarried.

Ores: typically an oxide of a metal, in the form of a rock.

Alloy: a mixture of two or more metals (or a metal with another element).

Ferrous: a material that contains iron.

Non-ferrous: a material that does not contain iron.

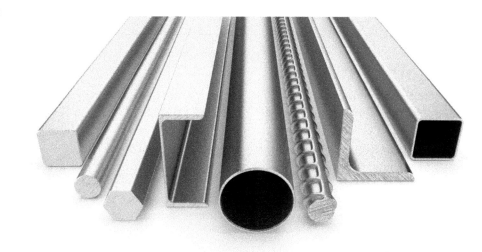

Figure 1.1.6 **Forms of stainless steel**

Non-ferrous metals and alloys

There are a wide range of metals and alloys that do not contain iron. Common non-ferrous metals and alloys include:

- aluminium and its alloys
- copper
- brass (an alloy of copper and zinc)
- bronze (an alloy of copper, tin and small amounts of other metals such as aluminium zinc, lead and silicon)
- lead
- zinc.

Figure 1.1.7 **Aluminium drinks cans**

Aluminium

Aluminium is one of the most commonly occurring elements on our planet. However, as a pure metal it is not as strong as steel, so it is normally alloyed to improve its properties. Compared to carbon steels, aluminium alloys cost more but have better resistance to corrosion. This makes them ideal for uses such as cans for soft drinks. Aluminium alloys also have lower density than steel; this means that aluminium products of the same size as steel products weigh much less, so aluminium alloys are also used for applications such as the wings and body panels for aircraft.

Copper, brass and bronze

Copper is an excellent conductor of electricity and is also ductile, so it is often used to make electrical wires. Unlike many other metals, it is commonly used in its pure form, as alloying reduces its ability to conduct electricity; however, copper oxide can be added to make it stronger. Copper also has very good corrosion resistance, so its other most common use is to make water pipes.

Copper is also used to make the alloys bronze and brass. Bronze is an alloy of copper and tin, which is often used for cast products, including statues. Brass is an alloy of copper and zinc. It is difficult to cast, but can be machined to a high finish. Its uses include pressure-valve bodies, doorknobs and musical instruments.

Figure 1.1.8 **Aluminium recycling symbol**

Figure 1.1.9 **Copper pipes in a boiler room**

Lead

Lead is relatively soft, malleable and ductile and has very good resistance to corrosion. For these reasons lead sheets have often been used in the construction industry, for example to prevent water leaks around the edges of roofs. It also has a high density, so it is used for applications such as weights for diving belts and shielding for radiation in nuclear reactors. It used to be commonly used for water pipes in houses, but it was found that in the long term, exposure to lead can cause health problems in humans; nowadays, other materials which are less hazardous to humans are used for water pipes.

Zinc

Zinc has a low melting point compared to most metals (approximately 420°C). As this means it does not need as much energy (and therefore cost) to melt it, it is commonly used for die casting. Products that are commonly made from die-cast zinc include handles for car doors and camera bodies. It can be alloyed with aluminium to increase its strength.

ACTIVITY

Metals are used in a wide range of products. Using a table similar to the one below, try to identify the metals found in common products.

What properties make the metal suitable for use in that product?

Product	Metal	Important properties
Bicycle frame		
Kettle body		
Door key		
Car door		
Spoon		

Changing the properties of metal products

Alloying is a common way of creating a metal with the properties needed by a product. Compared to a pure metal, an alloy may offer, for example, higher strength, toughness or corrosion resistance, depending upon the alloying elements used. It typically involves melting two or more metals together, so they become mixed at a chemical level. In the microstructure of the metal, it is not normally possible to make out the different pure metals.

There are other ways in which the properties of metal products can be altered. These include:
- modifying the structure of the metal
- changing the surface chemistry.

Modifying the structure of the metal

If a metal product is examined under a microscope, it can be seen to be made of lots of grains of material pressed together. The size and shape of these grains affects the mechanical properties of the metal. In general, as the grain size reduces, the metal becomes harder and stronger; however, it also becomes less ductile and more brittle. This grain size can be affected by either **cold working** or heat treatment.

Further, in some metals different grains can have atoms arranged in different ways; these different arrangements typically also have different mechanical properties. In some metals these arrangements, and therefore the properties, can be changed by heat treatment.

KEY WORDS

Cold working: repeatedly bending or hammering a metal.

Work hardening: an increase in the strength and hardness of a metal due to cold working.

Cold working

Many metals get harder as work is done to them – this is known as **work hardening**. It is why some metal parts get brittle and break after they have been repeatedly bent or hammered. When a metal is cold worked (i.e. not heated up before work is done to it), the grains in the affected area are deformed. They become stretched out, making them thinner, and effectively smaller, in that direction.

Further, within the metal grains there are many tiny flaws in how the atoms are arranged. These flaws are called dislocations. With repeated stressing, the atoms can move around within the grain, into the spaces left by these dislocations. In effect, the relocation of the atoms causes the dislocations to appear to move. However, when these moving dislocations meet up they can effectively 'pin' each other in place, stopping the atoms moving when stress is applied. This reduction in the ability of the atoms to move is what reduces the ductility and increases brittleness.

Heat treatments to modify grain size

If a metal is heated to a suitable temperature, the grains within it can grow. This makes the metal softer and easier to work. This process is called **annealing** and is explained in detail in Section 2.6. It is often used to soften metal that has been work hardened, or to make metals easier to bend into complicated shapes.

Hardening and quenching

High-carbon steel, containing 0.8–1.4 per cent carbon, can be **hardened** by heat treatment; low-carbon steels cannot be hardened in this way. After heating, the steel is **quenched** by cooling it rapidly, and then **tempered**. This results in a steel which has a combination of hardness and toughness. This process is described in detail in Section 2.6.

Normalising

Normalising is carried out on steel that has been work hardened. It results in steel that is tough with some ductility. This is described in detail in Section 2.6.

Changing the surface chemistry

The properties of the surface of a metal product can be affected by its environment. For example, it may be damaged by corrosion or mechanical forces. For metal products, it is often possible to alter the effect of their environment by either managing corrosion or, for steels, by changing the structure of the surface through the addition (or subtraction) of carbon.

Corrosion

Corrosion is where the surface of the metal reacts with another substance in its environment. For example, this could be aluminium reacting with oxygen in the air to form aluminium oxide, or low-carbon steel reacting with rain water to form rust.

With aluminium, the oxide layer that forms protects the metal against further corrosion. The layer is so thin that it is not normally visible to the naked eye.

With steel, the corrosion can be progressive – over time more rust forms, slowly eating away the thickness of the material. Even a small amount of rust can have an aesthetic effect on a product, which may reduce how attractive it is to a user. However, as corrosion increases this results in a reduction in the thickness of the metal. If the metal has a force applied to it, for example a tensile load pulling it, this means that there is less metal to resist the force, so the stress in the metal is higher. If enough material has been corroded away, the stress can exceed the yield strength of the material, leading to permanent deformation or even failure. To avoid this, designers normally try to reduce or prevent corrosion. This can be achieved by,

Figure 1.1.10
Metal failure after work hardening

Cracks in metal after bending

Figure 1.1.11 Corrosion on industrial pipes

for example, stopping the surface of the steel coming in contact with water, by:

● painting
● applying a plastic coat by spraying or dipping
● applying a layer of another metal that does not react with the water – this is carried out either by dipping the product in molten metal or by electroplating, which involves placing the product in a chemical bath
● attaching a metal that the water will react with rather than reacting with the steel – the material is sacrificed to protect the steel (for example, zinc blocks are often used as a sacrificial material for offshore applications, such as the legs of oil rigs or boat hulls).

Corrosion is not desirable as it reduces the effective life of metal products; materials are often chosen to avoid it. However, in some cases manufacturers will use materials knowing that they will corrode, limiting the usable life of the product. This may be on cost grounds, as a material is much cheaper and easier to process than a more corrosion-resistant alternative, or it may be a result of planned obsolescence by the engineer.

Addition or subtraction of carbon in steels

Some products require a combination of the toughness of the low-carbon steel, with the hardness of high-carbon steel. For example, the gears used in train engines need to be tough so that they do not fracture; however, they also need to be hard so that they do not wear out where the teeth touch each other. This can be achieved by making the gear from a tough low-carbon steel then increasing the amount of carbon in the surface. This produces a hard skin that will resist wear. This process is known as case hardening.

The case hardening process is made up of two parts: **carburising** and hardening.

Carburising involves adding carbon to the outer surface of the steel. For a single part in a small workshop, this can be done by heating the steel part to red hot and then dipping it into carbon powder. Some of the carbon powder will be adsorbed into the surface by diffusion. This will normally be repeated two or three times. In an industrial situation, particularly when manufacturing quantities of parts, the methods used are:

● The steel part can be packed in charcoal granules and then heated to a temperature of about 900°C. This is then held at temperature for a few hours, to allow it to 'soak', so that the carbon can diffuse into the surface.
● Gas carburising: the steel part can be heated in a special furnace, where the atmosphere is controlled. This will normally contain a known proportion of carbon-rich gas, such as carbon monoxide. This is then allowed to soak at temperature for several hours, so that the carbon can diffuse into the surface.

Typically, gas carburising gives the most accurate control over the percentage of carbon in the surface, and dipping in carbon powder is the least accurate method.

Following carburising, the metal product is normally heated to red hot and quenched. This involves rapid cooling by dipping it in water, brine (salt water) or oil, as described in Section 2.6. As there is only a very thin skin of hard metal, the centre of the steel product will remain soft; this means that tempering is not needed.

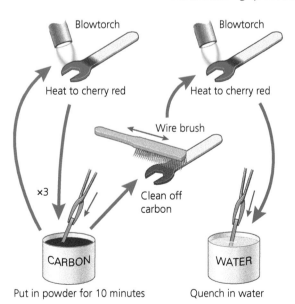

Figure 1.1.12 Case hardening of a low-carbon steel spanner

Available forms

Most metals are available in a wide range of standard forms and sizes:
- ingots (for melting to cast products)
- flat plates, sheets and strips
- bars and rods
- tubes and pipes
- standard section forms
- wire.

However, not all metals are available in every combination of size and shape, so engineers and manufacturers normally check what is available with their suppliers.

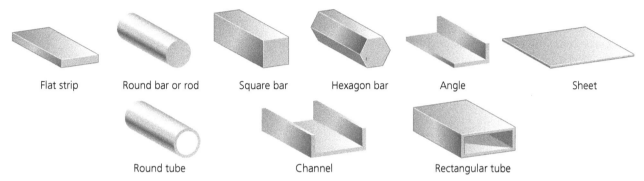

Flat strip	Round bar or rod	Square bar	Hexagon bar	Angle	Sheet

Round tube	Channel	Rectangular tube

Figure 1.1.13 Standard metal forms

Metal	Bar	Flat	Tube	Shaped sections	Wires
Low-carbon steel	Round, square or hexagon	Plate, sheet or strip	Round, square or rectangular	L section (angle), U channel, H section, Tee section	
Aluminium alloy	Round, square or hexagon	Plate, sheet or strip	Round, square or rectangular	L section (angle), U channel, Tee section	0.5–3 mm thick
Copper	Round or square	Sheet or strip	Round		0.5–3 mm thick
Brass	Round, square or hexagon	Sheet or strip	Round	L section (angle)	0.5–3 mm thick

Table 1.1.2 Commonly available metal forms

Form	Characteristic	Standard sizes (mm unless otherwise stated)
Sheet	Width and length:	1 m x 1 m, 2 m x 1 m
	Thickness:	0.6, 0.8, 1.0, 1.2, 1.5, 2.0, 2.5, 3.0
Strip	Width and thickness:	10 x 3, 25 x 3, 50 x 3, 12 x 5, 20 x 5, 50 x 5, 12 x 6, 20 x 6, 50 x 6, 50 x 25, 100 x 50
Bar	Diameter of round section or side of square section:	3, 4, 5, 6, 8, 10, 12, 16, 18, 20, 22, 25, 30, 35, 40, 45, 50

Table 1.1.3 Typical standard sizes for some low-carbon steel forms

Metals and alloys

Polymers

Polymers are the most widely used type of material in commercial production. They comprise a large number of similar, smaller chemical units that are bonded together. Most polymers are synthetic, which means that they are man-made using chemical processes. Synthetic polymers are typically made from crude oil, which is obtained by drilling underground or under the sea. This is then processed in a chemical plant. However, there are an increasing number of natural polymers, made by processing plants. These include latex from trees, which is used to make natural rubber, and corn starch polymers, which are increasingly being used for disposable food packaging and cutlery.

Figure 1.1.14 Offshore oil rig

There are two main types of polymer: **thermoplastics** and **thermosetting polymers**.

Thermoplastics

Thermoplastics consist of long chains of repeating chemical parts. The individual chains are only weakly attached to other chains. In some ways, a thermoplastic material resembles cooked spaghetti, in that the polymer chains overlap and entwine with each other to hold the material together. The weak links between the chains mean that thermoplastics are relatively ductile.

When thermoplastics are heated, they become softer and flexible. They can be shaped when hot and will harden into the new shape when cooled. If they are heated up again they can be reshaped. This property is a major reason that they are so widely used – it means that companies can buy sheets of polymers in standard sizes and easily change them into the shape required using heat, with processes such as vacuum forming or compression moulding. As well as being available in standard sheet sizes, they are also available as granules for use with moulding processes.

Figure 1.1.15 Water bottles formed by blow moulding thermoplastic tube

Thermoplastics can normally be recycled by melting them down. Many plastic products have markings to show the type of thermoplastic they are made from, to help them be sorted for recycling when they are thrown away.

Thermoplastic	Recycling symbol	Properties	Typical uses
Polyethylene terephthalate (PET)	1 PET	Clear, tough, shatter-resistant Good resistance to moisture	Drinks bottles, polyester fibres (polar fleece)
High-density polythene (HDPE)	2 HDPE	Hard, stiff Good chemical resistance Good impact strength	Bottles, buckets
Polyvinyl chloride (PVC)	3 PVC	Stiff, hard, tough, Good chemical and weather resistance	Window frames, guttering, pipes
Low-density polythene (LDPE)	4 LDPE	Tough, flexible, Electrical insulator Good chemical resistance	Detergent bottles, carrier bags
Polypropylene (PP)	5 PP	Hard Lightweight Good chemical resistance Good impact strength	Food containers, medical equipment
Polystyrene	6 PS	High Impact Polystyrene grades (HIPS) have good toughness and impact strength. Good for vacuum forming, injection moulding or extrusion.	Packaging, foam cups
ABS	9 ABS	Strong and rigid. Harder and tougher than polystyrene, but roughly twice the cost.	Plastic pipes, children's toys, keyboard keycaps
Acrylic	7 OTHER	Good optical properties – can be transparent. Hard wearing and will not shatter on impact.	Plastic windows, bath tubs, machine guards
Nylon	7 OTHER	Good resistance to wear. Low friction qualities. Ductile and durable.	Gear wheels, bearings
Polycarbonate	7 OTHER	High strength and toughness. Heat resistant. Excellent dimensional and colour stability.	Safety glasses, DVDs, exterior lighting fixtures

Table 1.1.4 Some common thermoplastics

Thermosetting polymers

In a thermosetting polymer, compared to a thermoplastic, there are extra links formed between the individual chains of polymer. These links stop the chains being able to move, and mean that thermosetting polymers are typically harder and more rigid than thermoplastics. However, once moulded they cannot normally be reshaped. When heated they do not melt or soften; they stay the same shape and eventually start to char or burn.

Thermosetting polymers are normally available as either liquids or granules that will form the polymer when mixed together. At the end of their usable life, these polymers cannot be recycled. Most thermosetting polymers typically end up in landfill.

Thermosetting polymer	Properties	Typical uses include:
Epoxy	High strength, stiff, brittle. Excellent temperature, chemical and electrical resistance.	Printed circuit boards, cast electrical insulators
Polyester resin	Good strength and stiffness but brittle. Very good temperature, chemical and electrical resistance. Lower cost than the other resins.	Bonding or encapsulation of other materials, suitcases/luggage
Melamine resin	Stiff, hard, strong. Resistant to some chemicals and stains.	Laminate coverings for kitchen worktops, impact-resistant plastic plates
Polyurethane	Hard with high strength. Flexible, tough and low thermal conductivity.	Foam insulation panels, hoses, surface coatings and sealants
Vulcanised rubber	Higher tensile strength, elastic. Resistant to abrasion and swelling.	Tyres, shoe soles, bouncing balls

Table 1.1.5 **Some common thermosetting polymers**

Figure 1.1.16 **Electrical sockets are normally made from thermosetting polymers**

Composites

Composites are materials made by combining two or more different types of material. For example, this could be ceramic fibres and a polymer, wood and an adhesive made from polymer, or metal and ceramic. Unlike a metal alloy, the materials are not joined chemically. If the structure of a composite is examined using a microscope, the different materials can be seen clearly, although they are physically next to each other.

Composites can have unique combinations of properties that are not possible in individual types of material, such as high strength and toughness with low density. The properties of a composite depend not just on the materials that it is made from, but also on the form of these materials and how they are distributed within the composite.

For example, glass-reinforced plastic (GRP), commonly known as fibreglass, is made from ceramic glass fibres in a thermosetting polymer. The fibres provide **reinforcement**, increasing the strength. The polymer creates a matrix around the fibres, holding them in place and making the material rigid. The greater the quantity of reinforcement, the higher the strength of the composite; however, this would also provide less matrix, so less rigidity.

Unfortunately, achieving the properties needed is not as simple as just changing the ratio of the reinforcement to matrix. If all the reinforcement fibres are aligned in the same direction, then the strength will be strongest in that direction. However, if a force was applied at a right angle to the direction of the fibres, the fibres would not increase the strength in that direction much, and the material could be pulled apart much more easily. For this reason, many composite products ensure that fibres are positioned either in layers pointing in different directions or woven together. This helps to ensure that properties are the same in different directions.

Fibres are not the only form of reinforcement used in composites. The reinforcement can be layers of material (called plies or laminates), as in plywood, or even just particles, as in medium density fibreboard (MDF). In each case, the size and shape of the reinforcement will have an effect on the properties of the composite.

Most composites are made using moulding processes, where their different 'ingredients' are added together to form the required shape. For example, for fibreglass this could be sheets of fibre that are positioned in a mould and then impregnated or soaked with a polymer resin. For plywood and MDF, wood laminates or particles are compressed with adhesive to form sheets. These sheets can subsequently be machined into the shape needed using woodworking tools.

As it is very difficult to separate the different material types within the composite, they cannot normally be recycled. At the end of their usable life, many are disposed to landfill.

Particle

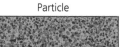

Laminated

Stranded

Figure 1.1.17 Types of composite

KEY WORDS

Composite: a type of material made by combining two or more different types of material. These remain physically distinct within its structure.

Reinforcement: the particles or fibre within a composite matrix that serve to increase its strength.

Figure 1.1.18 Sheets of carbon fibre, used to make composites

Composite	Properties	Typical uses include:
Carbon-fibre-reinforced polymer (CRP)	Carbon fibres in a resin matrix. Extremely high strength and rigidity. Low density. Expensive to produce.	Aircraft structures, high-performance sports bicycles, helmets
Glass-reinforced plastic (GRP)	Glass fibres in a resin matrix. High strength, low density. Good chemical resistance and thermal insulation. Lower cost than CRP but not as strong.	Canoes, small boat hulls, water tanks, surfboards
Plywood	Manufactured from layers of wood bonded together, at 90° to each other, using an adhesive. Smooth, even surface with good strength; often available in veneered form.	Furniture making, structural panelling; exterior grades used for boat building
Medium density fibreboard (MDF)	Manufactured from wood fibres and an adhesive matrix. Smooth, even surface with uniform properties. Easily machined and painted; often available in veneered form. Lower cost than plywood.	Furniture and internal panelling
Oriented strand board (OSB)	Manufactured from strands of wood compressed with adhesive matrix. Similar properties to plywood, but more uniform and lower cost.	Load bearing applications in construction, such as sheathing for walls and roof decking
Structural concrete	The most commonly used composite material. Concrete reinforced with steel bars to increase its tensile strength.	Bridges, high rise buildings

Table 1.1.6 **Some common composite materials**

Figure 1.1.19
Structure of plywood

Figure 1.1.20 **High-performance bicycle made from carbon fibre**

1.1 Materials and their properties

Other materials

In addition to metals, polymers and composites, structural grades of timber and ceramic materials are also used for engineering applications.

Timber

Timber is wood from trees. There are many types of wood available; however within GCSE Engineering consideration only needs to be given to structural grades. These may be used, for example, to make the frames for wooden houses or gliders.

Timber used for structural applications is typically softwood, meaning that it is produced from coniferous trees, which keep their leaves all year round. Examples include redwood (also known as Scots pine) and western red, cedar and spruce. These are usually sawn into standard sizes and shapes. Rough sawn timber may also be planed to give it a smooth surface. Planing makes the wood approximately 3 mm smaller on each face than the sawn size. Planed timber is more expensive than sawn timber. However, it has a smoother finish and a more accurate size.

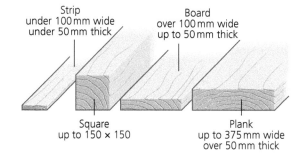

Figure 1.1.21 **Standard timber sections**

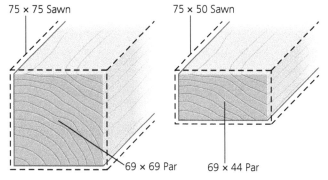

Figure 1.1.22 **Typical planed timber sizes**

Ceramics

The name **ceramic** comes from a Greek word meaning 'potter's clay'. However, ceramics are used for many more applications than just cups, plates and pots. In engineering their uses include:

- building materials, including concrete, bricks and plaster
- tools for cutting and grinding, made from tungsten carbide
- insulation for furnaces, made from alumina or aluminosilicates
- lenses, made from silicates.

Ceramic materials are typically oxides, nitrides or carbides of metals. They are usually harder than most other materials, meaning that they are very resistant to wear and scratches – hence their use for tools. They have excellent resistance to corrosion and are therefore used for chemical containers in laboratories.

Ceramics are very good insulators, for both electricity and heat, and can withstand high temperatures without softening. They also generally have good strength in compression. However, they have low tensile strength, very low ductility and are brittle. When they are subject to pulling forces they tend to crack and fall apart.

Due to their hardness, ceramics are very difficult to machine. They are often made by moulding processes, where very fine particles are either compressed together or held together by a liquid (like in clay). The moulded shapes are fired in furnaces at temperatures far above the melting point of most metals. This allows the ceramic particles to melt and join together. The use of moulding means that machining is not needed.

Figure 1.1.23 Stack of concrete masonry blocks

In theory, most ceramic materials could be recycled. However, due to the high temperatures required to melt them, this is not normally cost effective. The exception to this is glass, which is used in very high quantities for bottles; the enormous numbers of bottles used each day mean that it is possible to get economies of scale to recycle them. Most other ceramics end up in landfill at the end of their usable life.

KEY POINTS
- Materials properties include strength, ductility, malleability, hardness, toughness and stiffness.
- While categories of materials share some similar characteristics, each different material has a unique combination of properties.
- Ferrous metals contain iron; non-ferrous metals do not.
- The properties of metals can be changed by modifying the structure of the metal (such as by heat treatment) or changing the surface chemistry.
- Thermoplastics can be reshaped when heated; thermosetting polymers cannot be reshaped using heat.
- Composites are made from two or more materials, which remain chemically separate within the material.

Check your knowledge and understanding

1 State what is meant by the toughness of a material.
2 Explain the difference between a metal alloy and a composite material.
3 State three processes that involve the use of heat to change the properties of a low-carbon steel.
4 Explain why a manufacturer of saucepans might use handles made from a thermosetting polymer, rather than a thermoplastic.
5 Describe the typical properties of an engineering ceramic.

1.2 Material costs and supply

What will I learn?

By the end of this section you should have developed knowledge and understanding of:

→ how the availability of the form of material can affect the cost of a product
→ how the cost of a product includes contributions from both the materials it is made from and the processes used to make it
→ how to calculate the cost of materials in a product
→ the relative abilities of different materials to be processed.

When selecting a material for an application, it is a big advantage to an engineer if the preferred type of material is available in a variety of different forms. This allows the engineer to select a form that will minimise the amount of manufacturing needed, therefore keeping the cost down.

KEY WORD

Form: the shape and dimensions of a material.

Cost, availability, form and supply

Metals, polymers, composites, structural timber and ceramics are each available in a variety of different forms. These are described in the relevant section for the material type.

These standard **forms** and sizes are manufactured in large quantities by the companies that process the raw materials. This allows them to achieve economies of scale, keeping down the production costs. For example, a company may specialise in making square-section low-carbon steel tube. Their production line may produce several thousand metres of the tube each day, using dedicated, automated machines. They can then cut this into standard lengths, and sell it to the hundreds of companies that manufacture their own products using it. It is normally much cheaper for companies to buy these standard materials than to attempt to make the small quantity of materials that they need for their own products, as they would need lots of additional machines and skills.

Figure 1.2.1 Range of standard metal forms

MATHEMATICAL UNDERSTANDING

E3: Area of a circle

Question

A company has a rectangular sheet of material that is 1 m × 1 m in size. They need to cut out four circles each of diameter 0.4 m. After the circles have been cut out, the rest of the material will be waste.

a Calculate the area of one circle, to three significant figures.
b Calculate the percentage of material that will be wasted by this operation.

Solution

a Cross-sectional area $= \pi r^2 = \pi \times 0.2^2 = 0.126 \, m^2$
b Amount of material used $= 0.126 \, m^2 \times 4 = 0.504 \, m^2$
Amount of waste material $= 1 - 0.504 = 0.496 \, m^2$

Percentage of waste $= \dfrac{0.496}{1} \times \dfrac{100}{1} = 49.6\%$

Engineers will sometimes change their designs slightly so that a standard size can be used. For example, if 11 mm round bar is not available, the engineer may decide to use 10 or 12 mm diameter bar instead. The reason for this is that it is much more cost effective to buy material close to the size and shape needed for a product, rather than to buy a large lump of material and spend many hours machining it. It is also more cost effective to reduce the amount of waste material produced, as the cost for this material still contributes to the cost of the machined product.

Calculating costs

As well as the mechanical properties required by an application and **aesthetic** considerations, the engineer will normally have to consider the cost. This will include the cost of:

- the material. If a product must be manufactured from a non-standard form of material this may significantly increase the cost
- changing the form, or shape, of the material into that needed by the product.

For example, an engineer may compare the cost of machining a product from a standard-sized piece of aluminium alloy, to the cost of producing the product by moulding from carbon-fibre-reinforced polymer (CRP).

The engineer will probably also consider how easy it is to carry out the manufacturing operations needed. This may include, for example, whether the manufacturing company has the equipment needed and how complicated the manufacturing process will be. A complicated process may result in more errors, leading to scrap, waste and cost.

MATHEMATICAL UNDERSTANDING

Calculating costs (M14: use ratios, fractions and percentages)

Question
A company buys in materials and machines them to make a final product.
The cost of materials used in a product is £53.20.
The product requires a total of 2.5 hours machining, at a rate of £30 per hour (which includes labour and allowances for all other business costs).
a   Calculate the total cost of manufacturing the product.
b   The sales price of the product is £160.25.
    Calculate the percentage profit that the company makes on each product.

Solution
a   Cost of machining = 2.5×30 = £75
    Total manufacturing cost = cost of materials + cost of machining = 53.20 + 75.00
    $\qquad\qquad\qquad\qquad\qquad$ = £128.20
b   Profit per product = 160.25 − 128.20 = £32.05

$$\text{Percentage profit} = \frac{32.05}{160.25} \times \frac{100}{1} = 20\%$$

Machining, treating, shaping and recycling engineering materials

Table 1.2.1 compares a range of different materials. This is a subjective evaluation and will vary for different types of application. Note that:

- cost is based on cost per volume of material
- ability to be machined is how easily and effectively the form of the material can be changed by conventional machine processes
- ability to be treated is how easy it is to apply a finish to the surface of the material

- ability to be shaped is how easy it is to form the material into a shape when a product is first manufactured; this included both shaping and forming processes
- ability to be recycled indicates whether a material can normally be economically recycled.

Table 1.2.1 Subjective comparison of the cost and material characteristics of various materials

Material type	Material	Cost	Ability to be:			
			Machined	**Treated (finished)**	**Shaped (or formed)**	**Recycled**
Ferrous metal	Cast iron	Medium	Medium/low	Medium	Medium	Yes
	Low-carbon steel	Low	High	High	Medium	Yes
	High-carbon steel	Medium	Low	High	Medium/low	Yes
	Stainless steel	High	Medium	Not needed	Low	Yes
Non-ferrous metal	Aluminium	High	High	Not needed	Medium/high	Yes
	Copper	High	High	Not needed	Medium	Yes
	Brass	High	High	Not needed	Medium/high	Yes
	Bronze	High	High	Not needed	Medium/high	Yes
	Lead	High	High	Low	Medium	Yes
	Zinc	Medium	High	High	Medium/high	Yes
	Gold	Very high	High	Not needed	Medium	Yes
Thermoplastics	ABS	Medium	High	Not needed	Very high	Yes
	Acrylic	Medium	High	Not needed	Very high	Yes
	Nylon	Medium	High	Not needed	Very high	Yes
	Polycarbonate	Medium	High	Not needed	Very high	Yes
	Polystyrene	Low	Low	Not needed	Very high	Yes
Thermosetting polymers	Epoxy resin	Low	Low	Not needed	Medium/low	No
	Polyester resin	Low	Low	Not needed	Medium/low	No
	Melamine resin	Medium	Low	Not needed	Medium/low	No
	Polyurethane	Medium	Low	Not needed	Medium/low	No
	Vulcanised rubber	Medium	Very low	No	Low	No
Composites	Carbon-fibre-reinforced polymer (CRP)	Very high	Low	Low	Medium/low	No
	Glass-reinforced plastic (GRP)	Medium	Low	Medium	Medium/low	No
	Plywood	Low	Medium	Very high	Low	No
	Medium density fibreboard (MDF)	Very low	High	Very high	Low	No
	Oriented strand board (OSB)	Very low	Medium	Medium	Low	No
	Structural concrete	Low	Very low	Low	Medium/low	No
Timber (structural grades)	Redwood (Scots pine)	Medium	High	Very high	Low	No
Ceramics	Tungsten carbide	Very high	Very low	Very low	Very low	No

Examples of material selection

Figure 1.2.2 Electrical wires

Electrical wiring

Main considerations: electrical conductivity, ductility, cost.

The lowest cost metal would be low-carbon steel. However, copper has a much higher electrical conductivity and is more ductile, meaning that it can more easily be formed into wires; this makes it a better choice. In terms of electrical conductivity, gold would be an even better choice than copper, but due to its high cost it is not normally cost effective.

Figure 1.2.3 Scaffolding

Structural supports (scaffolding) for outside a building

Main considerations: strength, weight, cost, corrosion resistance.

Low-carbon steel scaffolding offers high strength. However, it is relatively heavy, which could increase the weight that the lower sections of the scaffolding must support. Structural timber weighs less than steel by volume, offers a high strength-to-weight ratio and, relative to steel, costs less. It is also easier to attach together (for example, with screws). However, structural timber may be damaged by rain when used outside. It can soften and rot when used outside, although this can be slowed down if it is treated with chemicals. While steel will rust, this happens much more slowly. Although steel scaffolding costs more than using wood, as it is less affected by rain and can be reused several times, it is typically the preferred choice.

Figure 1.2.4 Saucepans

Handles for saucepans

Main considerations: thermal insulation (resistance to heat), toughness, cost.

Metals could be used, as metal handles are easy to form into the required shape. However, the handle could get very hot when the pan is used, giving safety issues. Ceramics and polymers are both effective insulators, so would not get hot. Polymers are much tougher than ceramics, so would be less likely to break if accidentally knocked or dropped. Of the two types of polymer, thermoplastics may change shape and become weaker when they are heated, so are not suitable; this is not an issue for thermosetting polymers. Melamine would therefore be a suitable choice.

Figure 1.2.5 Cutting tool in a CNC machining centre

Cutting tools for machining centres

Main considerations: hardness, cost, toughness.

High-carbon steel and engineering ceramics are potential choices. Ceramics such as tungsten carbide are harder than the steel, offering better resistance to wear. However, the steel is tougher; ceramic tools can be brittle, with the risk that they will shatter if accidentally driven into the material being machined. It can also cost more to form ceramic materials into the shape required for the cutting tool. For high performance cutting of hard materials, ceramic tools may be preferred, but where toughness is required high-carbon steel may be the preferred choice. For less demanding applications, the lower cost of making high-carbon steel tools means that it is typically the preferred choice.

KEY POINTS

- It is normally much cheaper for companies to buy materials in standard sizes than to manufacture materials.
- The choice of material for a product can depend not only on the material properties, but also the forms in which the material is available.
- When comparing different materials for an application, these may need different manufacturing processes, each of which may have different effects on the overall cost of the product.

Check your knowledge and understanding

1 Explain why a company may buy in a standard size of material rather than making the material in the exact size needed.
2 Explain why the availability of material forms can result in changes to a product design.
3 Describe the factors that directly contribute to the cost of a product.

Factors influencing design of solutions

What will I learn?

By the end of this section you should have developed knowledge and understanding of:

→ the benefits and drawbacks of different methods of producing energy
→ the **engineered lifespan** of a product
→ how maintenance is used to extend the working life of products
→ other factors that can influence the choice of material and manufacturing process for a product.

In addition to the mechanical properties needed and the manufacturing considerations, other factors that can influence the design of a product include:

● energy requirements for production
● the engineered lifespan of the product, including any maintenance requirements
● availability of materials and user requirements.

Energy requirements

An important factor influencing the design of a product is the amount of energy needed to make it. This energy can have a significant cost, in terms of both money and the environmental impact.

Most of the energy used when making a product will go into getting a piece of material into the shape needed. Energy is used at every step of the process, for example:

● obtaining material – diesel fuel for the machines used to quarry metal ore
● refining material – electricity to run the chemical plant that refines oil to make polymers
● changing the shape of material – heat produced from electricity to melt plastic for injection moulding or gas flames used to melt metal for casting; or electricity to power the motors in machines
● changing the properties – gas or electricity used in heat treatment processes, such as hardening steel or curing composites
● transporting materials and products – fuel or electricity for vehicles.

Figure 1.3.1 Transporting materials and products

In addition, companies need energy to run some activities that do not directly involve making the product, such as heating, lighting and computer systems.

Most companies buy their energy from suppliers. Obviously, they need to buy the correct type of energy for their needs – lathes and milling machines with electric motors will only work when supplied with electricity. However, the energy providers can collect the energy from several different sources.

Sources of energy

Sources of energy can be classified as two types: **non-renewable** and **renewable**. Non-renewable sources are consumed when they are used and will eventually run out. For example, there is only a certain amount of crude oil on planet Earth, and as it is used it runs out. Using non-renewable energy sources is like taking petrol from a can – once you have used it all the can is empty and there is no more. Renewable energy sources are replaced naturally. For example, if you use the wind to power a generator, you cannot use it up – it will keep blowing. Renewable sources of energy are sustainable; they can be used again and again without running out.

Most energy sources, both renewable and non-renewable, generate electricity in the same way: they turn a generator or turbine. This may be either directly through their own movement, or indirectly by heating water to make steam, the pressure of which is used to turn the generator. The generator is effectively the opposite of an electric motor – rather than putting in electricity to turn the motor, the turning of the generator makes electricity.

Table 1.3.1 **Energy production methods**

Energy production method	Sustainability	Energy source
Fossil fuels	Non-renewable	Oil, gas and coal
Wind	Renewable	Wind (wind turbines)
Tidal	Renewable	The sea
Biomass	Renewable	Food oils
Solar	Renewable	The sun (solar panels)

Non-renewable energy sources

Fossil fuels

Fossil fuels include petrol and diesel (both made from crude oil), gas (either natural gas from underground pockets or shale gas) and coal. Traditionally these fuels have generated most of the world's electricity supply. In 2014 fossil fuels generated 66.7 per cent of the global electricity supply and accounted for 80.8 per cent of all energy consumption (source: World Bank).

Fossil fuels are burnt in large furnaces, which heat the steam to power the generators. Advantages of using fossil fuels are that:
- this is a reliable technology and energy can be produced as it is needed, responding quickly to changes in demand. For example, during the advert breaks on television there can be a sudden surge in demand when viewers get up to make a drink
- the power stations needed to produce the energy are already operational, so the need for new investment is limited.

However, using fossil fuels also has some serious disadvantages.
- The by-products of burning fossil fuels can cause significant damage to the environment. This includes pollution, smog, acid rain and gases that contribute to global warming.
- The price of fossil fuels is likely to increase as they are used up and become increasingly scarce.

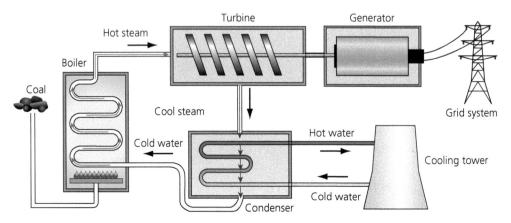

Figure 1.3.2 **How electricity is generated in a coal-powered power station**

Nuclear energy

Nuclear power uses certain types of radioactive materials, such as uranium. When enough of the radioactive material is put together, it generates heat due to its radioactivity. This collection of radioactive material is called a nuclear pile. The heat produced can be used to turn water into steam to power generators.

The advantages of nuclear energy are that, once set up, it can produce electricity at a low cost. Energy can be produced 24 hours a day, seven days a week, and it is also possible to respond to changes in demand to an extent. This is achieved by using carbon cooling rods, which are pushed into the nuclear pile to reduce the amount of heat that it produces.

However, the radioactive materials used in nuclear power plants can cause significant damage to both health and the environment, with effects lasting many years. Incorrect operation of a nuclear plant or a natural disaster can cause these radioactive materials to be released into the environment. This happened in 1986 at Chernobyl in Ukraine and, due to an earthquake, in 2007 at the Kashiwazaki-Kariwa Nuclear Power Plant in Japan. Due to the risks, nuclear plants must include many safety systems, which means they are very expensive to build.

Further, even when operating safely, nuclear power plants create radioactive waste; some of this is radioactive material that has been used in the nuclear pile, but it also includes other materials that have been exposed to radioactivity. Extreme care must be taken to dispose of these materials safely, as they can remain a health hazard for hundreds of years.

Figure 1.3.3 **Aerial view of a nuclear power plant on the California coast**

Renewable energy sources
Wind power

Wind power has been used for many centuries by windmills to grind cereal grains into flour. The modern use of wind power uses wind to generate electricity. Wind can turn the propeller blades of a wind turbine directly, although the rate at which the propeller turns is typically quite slow. The rotating propeller shaft is attached to a gearbox, which is itself attached to a generator. The gearbox increases the rate of rotation at the generator, allowing the turbine to generate electricity even when there is only a little wind.

The amount of energy collected is related to the size of the propeller blades and how quickly they rotate. Small wind turbines, with blades of less than 300 mm, can produce enough electricity to power small pieces of equipment. A large wind turbine based on land can have propeller blades of 10 metres or longer. A wind turbine based offshore, out at sea, can have blades that are more than 75 metres long. A 'wind farm' of six or more large wind turbines can generate enough electricity to satisfy the needs of several thousand homes.

Figure 1.3.4 **Wind turbine farm**

The main advantage of wind turbines is that once they are in place, electricity is produced at very low cost – effectively just the cost to maintain the equipment. The main disadvantages are:
- they only produce energy when there is some wind. This means that methods are needed to store the energy collected, which adds cost and reduces how efficient the approach is at providing energy
- the cost of building the wind turbine. However, compared to other methods of generating electricity this is relatively low
- some people think that it is not attractive to have large wind turbines where they can see them and dislike the noise that they make when working. While they want the electricity, they would prefer that it is generated somewhere else.

Tidal power

The flow of the tide at sea can be used to produce electricity in a similar way to wind power. The turbine blades can be located under the water and rotated by the movement of the water.

The advantages of **tidal power** are that:
- it is more consistent than wind power, as the tide is always moving
- it can be located where people are not concerned by what it looks like
- an offshore 'tidal energy farm' could be combined with a wind farm above the sea, maximising energy generation.

Figure 1.3.5 **Illustration of a tidal power farm**

The disadvantages of tidal power are that it:
- can be difficult to put into position and costs more to set up than wind power
- may have an adverse effect on other animals in the local ecosystem.

An alternative form of tidal power uses a chain of floats that are bobbed up and down by waves. While this has less effect on the environment, it generates much less electricity than the turbine blade system.

Biomass

Biomass and biofuel are organic material. This may be plants, vegetable oils, liquids fermented so that they contain alcohol, or even manure from animals. Biomass can be burnt, producing electricity in the same way as fossil fuels.

The advantages of biomass are that it is sustainable and that much of the biomass used for fuel would otherwise be waste, ending up in landfill or incinerated. The disadvantages are that weight for weight, biomass is less efficient as a means of generating energy than fossil fuels. Further, the by-products of burning biomass include some gases which contribute to global warming.

Solar power

Solar power uses sunlight to generate electricity. It is different from the other energy sources in that it does not involve turning a generator. It uses photovoltaic cells, where the light falling on the cell causes it to generate electricity. The amount of electricity generated is in proportion to the area of the cells used.

Figure 1.3.6 **Solar energy panels on a house**

The main disadvantage of solar energy is that it requires sunlight, so it cannot generate electricity at night. This means that methods are needed to store the energy collected, which adds cost and reduces how efficient the approach is at providing energy. The other disadvantage is that a large area is needed to generate the power required for even one home – even if solar panels cover the roof of a house, they will not generate enough energy to meet the full needs of a typical home, although they will make a significant contribution. A 'solar energy farm' may need to cover the area of several sports fields, which means that the cost of land can be an important factor. Further, the land cannot be used for other processes, such as housing, farming or industry. However, the efficiency of generating solar electricity is improving as the technology develops.

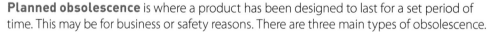

Engineered lifespans

Most products have a limited lifetime. This may be due to a deterioration in the product's ability to meet its requirements, or it may be deliberate, due to design. If it is due to deterioration, this can often be addressed by putting in place actions to increase the product life.

Planned obsolescence is where a product has been designed to last for a set period of time. This may be for business or safety reasons. There are three main types of obsolescence.
- Quality: the product is designed to wear out or break down.
- Function: the existing product becomes out of date because new, improved products become available.
- Desirability: the product is designed to go out of fashion.

Figure 1.3.7 **Mobile phones go out of date as improved products become available**

Planned quality obsolescence could involve, for example, selection of materials so that they will wear out. This requires careful calculation and understanding of the properties, otherwise the product could fail disastrously early. Alternatively, it could involve the use of sealed parts in the product. These are parts enclosed in compartments that cannot easily be opened; for example, they could be sealed by welding or using adhesives. If one of the parts breaks or wears out, typically the whole product would need to be replaced. This is often carried out as the parts cannot be replaced by the user or for safety reasons.

The main advantage of planned obsolescence is that the manufacturer might be able to sell more products: once the product is no longer usable, the user will have to buy a replacement. It may also reduce the need for maintenance in service, as the user may not have to take actions to extend the life of the product. The main disadvantage is that it places extra demands on the environment – additional resources are needed to make the replacement products.

> **ACTIVITY**
>
> Identify three products you use every day that have a limited life. What materials or features show you that these products are designed for a limited life? What is the reason for this planned obsolescence?

Maintenance of engineered products

Maintenance means carrying out actions to extend the life of a product. This is important to ensure that the product continues to work efficiently and continues to be safe in operation.

There are two types of maintenance: reactive and proactive.
- **Reactive maintenance** involves repairing broken parts.
- **Proactive maintenance**, also known as planned maintenance, involves carrying out actions that will prevent the product failing in the first place.

In general, proactive maintenance is preferred to reactive maintenance. This is because there is no unexpected stoppage in the function of the product. This is why, for example, cars are serviced at set intervals.

When considering an individual product, proactive maintenance may include several actions.
- **Lubrication**: this helps to reduce wear between parts, such as gear wheels. It also reduces friction between parts, which could otherwise result in parts heating up. This could cause thermal expansion that increases wear and causes parts to jam or seize together.
- Avoiding corrosion: corrosion can slowly consume an affected material. This reduces the amount of material, effectively reducing the mechanical properties of the part. In extreme cases, it may even cause holes and leaks in a casing. Lubrication may provide some protection against corrosion, for example, by providing a thin layer of protective oil on a part. Alternatively, other protective coatings may be applied, such as regularly repainting a product, to ensure that it is always protected from any potential sources of corrosion. This may also include regular cleaning, to remove materials or chemicals that may cause corrosion.
- Compensating for wear: in a complicated product with moving parts, it may be possible to adjust the position of some parts to allow for wear. Alternatively, parts that are starting to show signs of wear may be replaced before they fail; for example, the batteries in a controller may be replaced when their charge falls to a certain level, so that the controller never fails.

One method that is used to identify when proactive maintenance is needed is condition monitoring. This involves the use of statistics to predict the expected lifetime of components. For example, the temperature of a machine may be monitored and seen to slowly increase over time. When the temperature reaches a certain level, this could indicate that the lubricant needs to be topped up and replaced.

KEY WORDS

Engineered lifespan: the amount of time that a product is designed to last.

Planned obsolescence: designing a product so that it will have a limited lifespan.

Maintenance: activities which extend the life of a product.

Reactive maintenance: repairing broken parts.

Proactive maintenance: carrying out actions that will prevent a product failing.

Lubrication: using a fluid or other substance to reduce the friction and wear between contacting parts.

Figure 1.3.8 Adding lubrication to gears

STRETCH AND CHALLENGE

Select one of the machines that you have used in your school workshop. If possible, find the manufacturer's handbook. Think of all the different maintenance activities that might need to be carried out. What parts might wear out and need to be replaced? Where might lubrication be needed?

+ −
× =

MATHEMATICAL UNDERSTANDING

M13: Perform calculations using time and cost

Question

An engineering company collected data on how long cutting tools lasted before breaking on one of their machines. The results are shown in the table. Data was recorded for 50 tools. All of the tools failed in less than 60 hours.

Hours to failure	0 < 10	10 < 20	20 < 30	30 < 40	40 < 50	50 < 60
Number of failed tools	1	5	12	12	14	6

a   Produce a graph of the cumulative number of failures against the time to failure.
b   Calculate the mean average number of hours until tool failure.
c   To reduce unplanned stoppages, the company have decided to change tools after a set time, before the average tool fails. They have decided that the operational life of the tool should be set at the value when 30 per cent of the tools failed. Determine this value for the operational life of the tool.

Solution

a

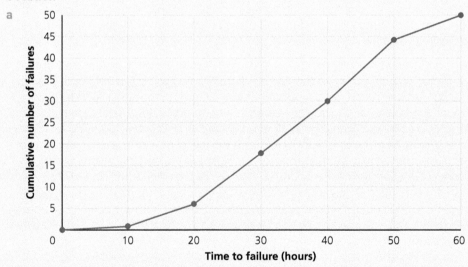

b   Average value $= \dfrac{((1 \times 5) + (5 \times 15) + (12 \times 25) + (12 \times 35) + (14 \times 45) + (6 \times 55))}{50}$

   $= 35.2 \, \text{hours}$

c   30% failure equates to 15 (out of the 50) components.
   From the graph, the operational tool life is therefore approximately 27 hours.

Availability of materials and user requirements

Sometimes other factors outside of the engineer's control mean that optimum combination of material and manufacturing process may not be possible or acceptable.

Changes in availability may mean that the material form selected is no longer available. This can arise, for example, due to:

● suppliers going out of business
● market forces, where demand for a material is greater than supply
● limits on the ability to obtain a material from a foreign country, such as local wars or trade restrictions.

This may mean that a non-standard form of material must be used, potentially increasing both the material cost and the cost of manufacture.

The end user of the product may also decide to give requirements additional to the properties needed to fulfil the function, such as wanting materials with higher properties or a specific 'look' to be used. For example, the end user of a frame for a racing bicycle may specify that it be made from a titanium alloy or a carbon fibre composite, so that it has higher strength and lower weight. This could also affect the manufacturing, particularly if a material requires a specialist manufacturing process.

KEY POINTS
● Electricity can be generated using renewable and non-renewable sources.
● Non-renewable sources, such as fossil fuels, will eventually run out. They also contribute to environmental pollution and global warming.
● Renewable sources are sustainable. However, some, such as solar or wind power, may not be able to produce a continuous supply of electricity. This means that they need additional energy storage facilities.
● Some products are designed to last for a limited period of time. This planned obsolescence may be for business reasons, to sell replacements, or for safety reasons.
● Maintenance extends the usable life of a product. It can be reactive or proactive.

Check your knowledge and understanding

1 Explain how the use of fossil fuels can affect the environment.
2 Describe how electricity is generated using nuclear power.
3 Give one advantage and one disadvantage of using wind to generate electricity.
4 Give three examples of maintenance activities.

PRACTICE QUESTIONS: engineering materials

1 Which of the following properties is defined as 'the ability of a material to resist bending'?
   A Strength
   B Malleability
   C Toughness
   D Stiffness

2 Which of the following means malleability?
   A The ability of a material to resist bending
   B The ability of a material to be deformed without rupturing
   C The ability of a material to resist an applied force
   D The amount that a material can be deformed

3 In a tensile test, the maximum force applied to a test piece before failure was 17 000 newtons. The cross-sectional area of the test piece was 85 mm^2. Calculate the maximum stress within the material. Show your working.

4 Explain the difference between a ferrous metal and a non-ferrous metal.

5 Explain why zinc is often used for die casting, rather than steel.

6 Explain why a metal may be 'normalised' by heat treatment.

7 Describe three methods used to stop steel products corroding in water.

8 List four forms in which metals are commonly available.

9 Explain the difference between the typical properties of thermoplastics and thermosetting polymers.

10 Explain how the quantity and orientation of the reinforcement used affects the properties of a composite material.

11 List three composite materials. For each, give a typical application.

12 Explain why ceramics (such as tungsten carbide) are used to make cutting tools for machining centres.

13 Describe how electricity is generated using wind power.

14 Explain how the end user can influence the designer's choice of materials for a product.

Engineering manufacturing processes

Manufacturing processes are operations that change the form or properties of a material. This means that they might be used to change the shape, size, composition or material structure in some useful way.

Processes are typically classified by the way in which they change the material. For each process type, there may be several different tools or pieces of equipment that could carry it out. These tools may also have different levels of automation ranging from manual tools to machine tools, to fully automated or robotic equipment.

This section will cover the different processes and equipment used in the manufacture of parts, such as:
- the type of material that the part is made from – some processes or tools are only suitable for use with certain types of material
- the forms in which the material is available – such as sheets, bars, granules, etc.
- the quantity of parts to be made – it is not normally cost effective to buy a special machine if only one product is to be made
- the cost of both the equipment and the labour needed to make the product.

This section includes the following topics, representing different types of manufacturing process:

2.1   Additive manufacturing

2.2   Material removal

2.3   Shaping, forming and manipulation

2.4   Casting and moulding

2.5   Joining and assembly

2.6   Heat and chemical treatment

2.7   Surface finishing

At the end of this section you will find practice questions relating to engineering manufacturing processes.

2.1 Additive manufacturing

What will I learn?

By the end of this section you should have developed knowledge and understanding of:
→ how metal products are made by sintering
→ rapid prototyping processes used with polymers
→ how fused deposition is used to make products.

Currently, the most common way of manufacturing a component in an engineering workshop is to start with a large piece of material and remove the material that is not needed. The end result is a smaller piece of material in the shape that is required and a quantity of removed 'waste' material. **Additive manufacturing** is different – rather than taking material away, it involves building up a part by adding material where it is required.

Sintering

Sintering is a process that is used to make products from metal powders.

A mould is made of the required product which is filled with metal powder. High pressure is applied to the powder using an industrial press. The powder is also heated to a temperature below the melting point of the metal. The combination of the pressure and heat makes the individual particles of metal powder fuse together at the points where they touch. In general, the higher the pressure and temperature, the greater the strength of the final product.

A major advantage of sintering is that it can be used to create solid metal products at temperatures below the melting points of the metals used. This means that the sintering process does not need as much energy as the casting process, which involves melting the metal.

Sintering can also be used to create metal products from different mixtures of powders, where the metals would not normally mix together effectively if they were molten. It can also create products with lower densities than solid metal, by allowing some air gaps to remain in the material structure.

Rapid prototyping for polymers

Rapid prototyping involves using additive manufacturing to make a complete part or component in a single operation. There are several different rapid prototyping processes used to make parts from polymers, including fused deposition modelling and stereolithography.

One advantage of rapid prototyping is that, compared to conventional workshop processes, a product is made in a single operation from a 3D computer aided design (CAD) drawing (see Section 6.3 for more detail). For complex parts, this can be carried out more quickly than the several conventional machining operations needed to make the part. This allows for rapid evaluation of whether, for example, a product may fit in the available space (for example, a turbocharger in an engine compartment).

> **KEY WORDS**
>
> **Additive manufacturing**: a manufacturing method where a part is built up by adding material where it is required.
>
> **Sintering**: the use of heat to convert a powder into a solid, without becoming a liquid.
>
> **Rapid prototyping**: the use of an additive manufacturing method to make a complete part or component in a single operation.

A disadvantage of rapid prototyping is the limited range of materials that can be used. The most common processes can be used only with certain types of polymer; other processes use flour and adhesive or even layers of paper. Rapid prototyping processes for metals are not currently commercially available, although they are under development. This means that if a prototype product is being made to test out a design, it may not have the same mechanical properties as the material that will be used in the final product.

A unique feature of rapid prototyping processes is that they can make products with internal features that cannot be made by conventional machining processes. This means unique designs can be made. However, it can also be a limitation if the part being made is a design prototype, as it may not be possible to make it in the final material using conventional **manufacturing processes**.

Fused deposition modelling

Fused deposition modelling is the most common form of rapid prototyping process. This is the process used in 3D printers. It is used to make products from polymers such as ABS, PLA, polyamides and nylon. The steps involved in the process are:

- a 3D CAD drawing is created of the part, which has been broken down into a series of layers
- the first layer of polymer is melted and deposited by the printing head, so that it sits on the base. The temperature at which the polymer melts is typically between 200 and 250 °C and it quickly cools and solidifies when it is deposited
- the printing head then rises by the thickness of one layer and deposits the next layer of polymer, building on the first layer
- this process of depositing a layer and rising continues until the full thickness of the product has been deposited
- the finished item can then be removed from the base.

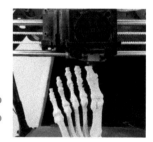

Figure 2.1.1 **3D printing model of bones in a human foot**

Stereolithography

Stereolithography is a rapid prototyping process that uses a laser to make polymer products (see Figure 2.1.2).

The process starts with a 3D CAD drawing and computer software is used to break down the drawing into a series of layers (saved in a format called an STL file). Each layer may be only a few microns in thickness. A movable platform is placed in a tank of **resin**, which is a liquid form of polymer. The laser then creates the product layer by layer.

First, the laser draws the shape of the first layer in the resin, so that it sits on the movable plate. The laser draws wherever the material is needed, causing the resin to turn into a hard polymer in those places.

The movable plate is then lowered by the thickness of one layer. The laser then draws the second layer in the resin, so that it builds on the first layer.

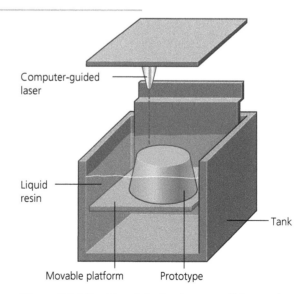

Computer-guided laser

Liquid resin

Movable platform

Prototype

Tank

Figure 2.1.2 **Rapid prototyping by stereolithography**

This process of drawing and lowering continues until every layer has been drawn, and the full thickness of the product has been created.

The final part can then be raised above the resin bath and any unhardened resin allowed to drain away.

ACTIVITY

Prepare a short article that could be used on a radio news show or podcast, explaining how rapid prototyping may change the way we manufacture products in the future, and how this may affect our lives.

KEY POINTS
- Additive manufacturing processes make products by adding material where it is needed.
- Sintering uses heat and pressure to make solid parts from metal powder.
- Rapid prototyping uses additive manufacturing methods to make a product in a single operation.
- The starting point for a rapid prototype is a 3D CAD drawing which is divided into a series of layers by software.
- Fused deposition modelling builds up a part using polymer deposited by a printing head.
- Stereolithography builds up a part using polymer from a resin bath that is hardened by a laser.

Check your knowledge and understanding

1 Explain what is meant by additive manufacturing.
2 State two factors that influence the strength of a metal part made by sintering.
3 Name two materials that can be used to make a part by fused deposition modelling.
4 Describe the process of making a part using stereolithography.

2.2 Material removal

What will I learn?

By the end of this section you should have developed knowledge and understanding of:

→ the tools and machines used to cut material by sawing, shearing and laser cutting
→ the uses of a lathe including turning and boring
→ the uses of a mill including facing and slotting
→ types of drill
→ how chemical etching is used to manufacture printed circuit boards.

Materials removal is sometimes called **wasting**. This starts with a piece of material larger than the part being manufactured and involves using tools or machines to remove the material that is not needed in the required form of the part. There are many different ways to remove the excess material, including:

- cutting
- turning
- milling
- drilling
- chemical etching.

Cutting

Cutting is used to reduce the length of a material or to remove a section of material. The different types of cutting process include:

- sawing
- shearing
- laser cutting.

Other cutting processes used in industry include thermal cutting, where a flame is used to melt through material; and abrasive water jet cutting, where material is worn away by a high-pressure jet of water containing particles of an abrasive material.

Sawing

Saws use movement to progressively cut away material. Each saw tooth will cut a small groove into a material as it moves against it. Saw teeth are normally angled slightly out from the blade. This means that the cut groove is slightly wider than the blade, so the blade is less likely to get stuck. The debris removed by the cut is typically deposited as the saw emerges from the other side of the material. The saw can only cut as fast as the debris can be removed; applying extra force to the saw will not make it cut faster.

Different types of saw are used to cut different materials and saws can have different sizes of teeth, so that they can cut different types and thicknesses of material. Typically, the harder the material, the smaller the saw teeth.

Hacksaw

Hacksaws have fine teeth and are used to cut metal and polymers. Junior hacksaws also have fine teeth and are used to cut thinner sections of the same material. Mechanical saws are like large hacksaws. They are used to cut metal bars and rods.

Coping saw

The narrow blade of a coping saw means that it can be used to cut curved shapes in thin pieces of polymer or timber.

Tenon saw and rip saw

Straight cuts can be made in structural timber using a tenon saw or a rip saw (also known as a sheet saw). These have much larger teeth than the saws used to cut metal.

Jigsaw

Jigsaws use a motor to move the saw blade. A wide variety of jigsaw blades are available, allowing them to be used for most types of material and thicknesses.

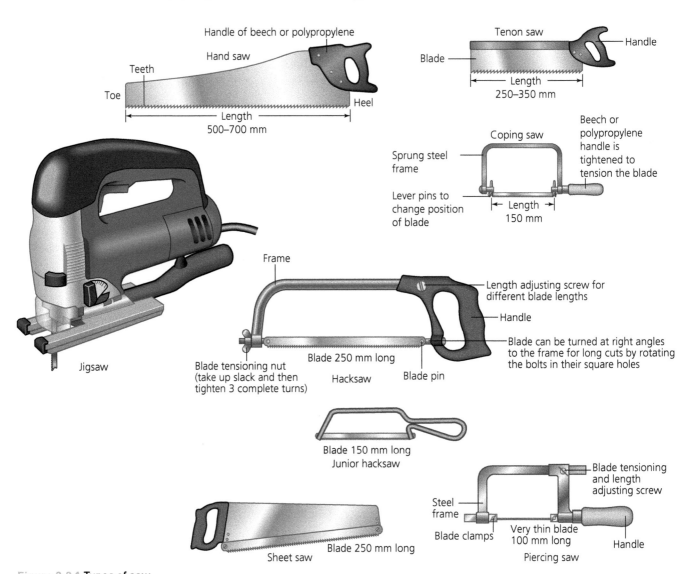

Figure 2.2.1 **Types of saw**

Shearing

Shearing involves applying a force from opposite sides of a metal sheet (Figure 2.2.2). This causes the sheet to separate in line between the two points where the force is applied. The amount of force needed depends upon the strength and thickness of the material.

Tin snips and shears are used to cut thin sheet. **Guillotines** are used to cut thicker sheet. Industrial guillotines may apply a force of hundreds of tons to push a blade made of tool steel through the material.

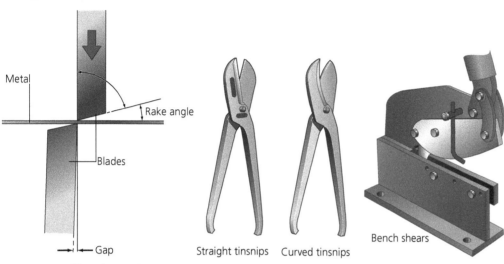

Figure 2.2.2 **Shearing action**

Figure 2.2.3 **Shearing equipment**

Laser cutting

Laser cutting is used to cut thin sheets of polymer or metal. The material along the cut line is vaporised. For health and safety, the vapour is normally sucked from the cutting machine using local exhaust ventilation.

This process must be controlled by a computer using either a written program or a CAD drawing, as it is not safe to control this equipment by hand. Although the equipment required for laser cutting is relatively expensive compared to the other cutting methods, it is very flexible. It is often used to accurately make small quantities of parts.

Figure 2.2.4 **Industrial laser cutter**

Turning

Turning involves the use of a **lathe** to make parts, or features on parts, with a round profile. The material to be machined, known as the 'workpiece', is held by the lathe and rotated at high speed. The cutting tool is pressed into the workpiece to remove the material.

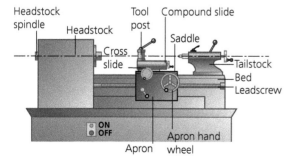

Figure 2.2.5 **Centre lathe**

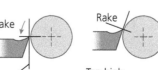

Figure 2.2.6 **Setting lathe tools to the correct height**

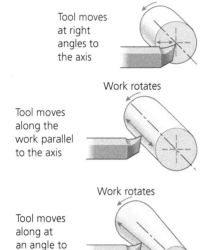

Facing off

Work rotates
Tool moves at right angles to the axis
Facing off

Work rotates
Tool moves along the work parallel to the axis
Parallel turning

Work rotates
Tool moves along at an angle to the axis
Taper turning

Parting off

Figure 2.2.7 **Lathe operations**

Facing off

The lathe can also be used to make a flat **face**, for example, creating a step at some point along the length of a round part or on the end of a turned part. This process is called 'facing off'. The face produced must be at right angles to the axis of rotation of the workpiece.

Cylindrical turning

Cylindrical turning produces a uniform shape of the same diameter along the turned length. It is also known as 'parallel turning'.

Taper turning

In taper turning, the diameter is larger at one end of the turned section, and progressively reduces towards the other end producing a tapered shape.

Drilling and boring

Turning can also be used to make holes in or through round workpieces. The workpiece is rotated as before, but here the cutting tool is a drill bit that is pushed into the end to make the hole. This is known as '**centre drilling**'.

Boring is the use of a single-point tool to produce an internal hole. This can be used for a variety of purposes, including making small adjustments to internal diameters, producing internal tapers and steps, or cutting an internal thread in a hole.

Right-hand knife

Left-hand knife

Round nose

Parting

Figure 2.2.8 **Typical turning tools**

Figure 2.2.9 **Part mounted in an industrial metal lathe**

CNC lathes are increasingly used to perform turning operations. CNC lathes are controlled by a computer, which is programmed to perform different operations. A CNC lathe can be fitted with a wide range of tools that allow it to taper turn, face off, drill, part off and parallel turn.

The solid tools shown in Figure 2.2.8 are increasingly being replaced with insert bits.

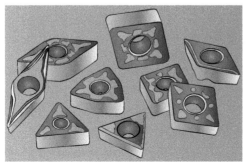

Figure 2.2.10 **Insert bits and a CNC lathe**

2.2 Material removal

M1.6: Calculate angles of a triangle using trigonometry

Question

A piece of material is to be turned on a lathe. Over a length of 96 mm, the material will have a reduction in diameter of 12 mm. Calculate the angle of the taper, shown as θ on Figure 2.2.10.

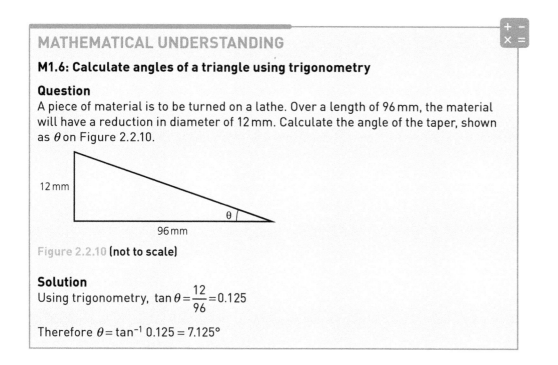

Figure 2.2.10 **(not to scale)**

Solution

Using trigonometry, $\tan\theta = \dfrac{12}{96} = 0.125$

Therefore $\theta = \tan^{-1} 0.125 = 7.125°$

Milling

Milling machines use a rotating tool to remove metal one thin layer at a time. They can be used to face a piece of material, producing a flat surface with a good finish. They can also be used to create slots in material. A **slot** can be a square groove open at the sides of the material or, alternatively, it can be made by 'plunging' the cutting tool into the material and moving it along, creating a circular shape at the end of the slot.

There are two types of milling machine: horizontal and vertical. This name refers to the axis about which the milling tool rotates.

> **KEY WORD**
>
> **Slot**: a groove in a material.

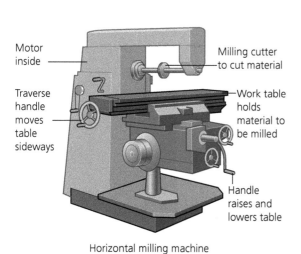

Motor inside

Traverse handle moves table sideways

Milling cutter to cut material

Work table holds material to be milled

Handle raises and lowers table

Horizontal milling machine

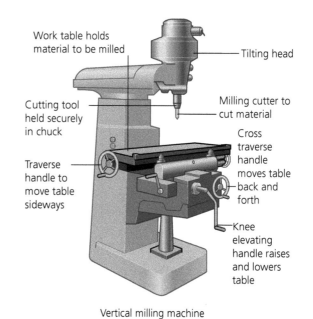

Work table holds material to be milled

Cutting tool held securely in chuck

Traverse handle to move table sideways

Tilting head

Milling cutter to cut material

Cross traverse handle moves table back and forth

Knee elevating handle raises and lowers table

Vertical milling machine

Figure 2.2.11 **Milling machines**

Milling of PCBs

Figure 2.2.12 Milling a component

One particular application of milling is to manufacture printed circuit boards (PCBs). PCBs can also be made using a small computer-controlled milling machine or router. This uses a CAD drawing of the PCB layout and a strip of insulating material with a thin layer of copper. The milling tool cuts a shallow groove through the copper layer around the outside of the tracks in the PCB design. Compared to chemical etching (explained below), an advantage of this process is that it does not use hazardous chemicals which could harm the environment. A limitation is that the separation between the copper tracks must typically be slightly larger than those made by chemical etching, to allow the tool access to cut away the unwanted copper. This means that the PCB is typically slightly larger.

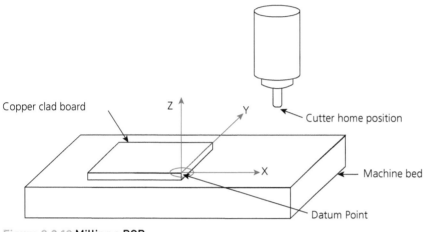

Figure 2.2.13 Milling a PCB

Figure 2.2.14 Making holes in a PCB using a CNC drill

Drilling

Drilling makes holes in the material using a rotating tool to progressively remove material. The exception to this is centre drilling, where the drill remains stationary and the workpiece is rotated in a lathe. This was explained above in the section on turning.

> **KEY WORD**
>
> **Pillar drill:** a type of drill mounted on a pillar or stand.

Traditionally, drilling was carried out manually using a hand drill. Portable or battery powered drills use electricity to reduce the human effort required. If the part being drilled can be moved, a **pillar drill** can be used. This gets its name from the pillar that the motor/tool and machine bed are attached to. It is important during the actual process of drilling that the part being drilled is firmly held using either a vice, a jig or a g-cramp. Many injuries are caused when (unheld) parts spin round during drilling.

Different types of drill are used for different materials. The speed that the drill needs to rotate for efficient cutting depends upon the material.

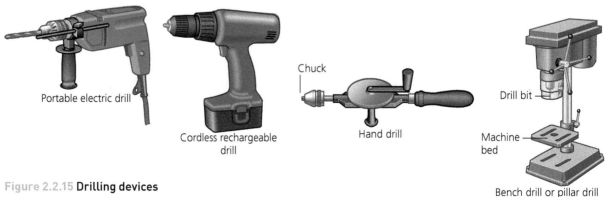

Figure 2.2.15 Drilling devices

Chemical etching

Chemical etching uses chemicals to remove material, rather than a tool. This method is often used to make PCBs. The PCBs are made from a strip of insulating material that is coated with a thin layer of copper.

Photo-etching

Photo-etching is a type of chemical etching used to make PCBs. It uses PCB material coated with a light-sensitive film on top of the copper. Photo-etching follows these steps.

- A mask is made, which is typically a piece of transparent material, with the PCB design printed on to it.
- The mask is placed over the light-sensitive film.
- The light-sensitive film is exposed to ultraviolet light for a set period of time. The mask protects the PCB design from the light.
- The mask is removed and the material placed in a bath of developer, to make the image of the PCB design appear on the board.
- The material is placed in a chemical bath. This typically contains ferric chloride, which eats away any areas of the film that were exposed to light, along with the thin layer of copper in these areas.
- The PCB is rinsed in water to remove the chemicals, stopping the etching process.
- Any holes needed to insert components are drilled into the PCB.

KEY POINTS

- Saws use movement to progressively cut away material.
- Tin snips, shears and guillotines cut materials using a shearing action.
- Lathes are used to produce round profiles on parts. The external profile may be cylindrical or tapered, and internal holes can be centre drilled or bored.
- Milling machines remove metal one thin layer at a time. They are used to produce flat surfaces and slots.
- Hand drills, portable and cordless drills and pillar drills can be used to make holes in material.
- Chemicals can be used to etch away unwanted metal from copper-coated board, leaving the printed circuit board design.

Check your knowledge and understanding

1. Name two types of saw used to cut metal.
2. Describe how a laser produces a cut in a sheet of polymer.
3. Describe how the outside of a part is machined using a lathe.
4. Explain the difference between the parts machined by turning and milling.
5. Other than a lathe, name two types of drill.

What will I learn?

By the end of this section you should have developed knowledge and understanding of:

→ the tools and equipment used to bend and fold sheets of material
→ how press forming, punching and stamping are used to manufacture parts
→ how composite materials are manufactured using the lay-up method.

KEY WORDS

Forming: a type of process that changes the dimensions or shape of a solid material, without changing the volume.

Shaping: a type of process that involves pouring or forcing liquid material into a mould.

Bending: forcing something straight into a curve or angle.

Former: a device in a required profile that a material can be formed against.

Jig: a device to hold a workpiece.

Forming processes change the dimensions or shape of a solid material, normally by applying force. These include bending, folding, and press forming. Typically, this is achieved by applying force to manipulate the material. The force may be applied by, for example, hammering, forging, mechanically pressing, air pressure or a vacuum. Some materials are easier to form if they are heated before forming.

Shaping involves pouring or forcing liquid material into moulds, and allowing it to set or harden to take on the shape of the mould. This allows complex 3D shapes to be created in a single operation. Shaping processes include most forms of casting, injection moulding and composite lay up.

Bending

Bending involves physically deforming a material. The material to be bent must be ductile and malleable; brittle materials tend to shatter when bent.

Thin sections of metal are often bent without heat, even by hand, though they are easier to bend when heated, especially for thicker sections. Non-ferrous metals such as aluminium and copper are often annealed before bending (see Section 2.6), to make them more malleable.

If an accurate bend or a particular shape is needed, then a **former** is often used. When a quantity of parts need to be bent accurately to the same shape, bending **jigs** are frequently used. Jigs are devices that hold the workpiece or material, and bending jigs are a type of jig that allow specific shapes or sizes to be achieved when bending a material. Where very large quantities of parts are needed, special-purpose bending machines may be used. These often use hydraulic pressure to apply a large force, forcing the metal strip around formers.

Thermoplastic polymers can be readily bent when heated, but most will snap or break if bent when at room temperature or below. A strip heater can be used to heat just the area where a bend is required. Once flexible the polymer can be bent but it must be held in position until it cools and hardens in the new shape as it may sag or deform.

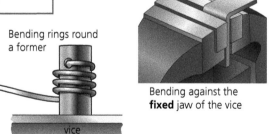

Bending rings round a former

vice

Bending against the **fixed** jaw of the vice

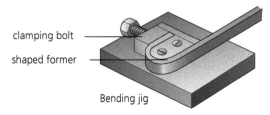

clamping bolt

shaped former

Bending jig

Figure 2.3.1 Examples of metal bending

Folding

Folding is bending material over on itself, so that one part of it covers another. Folds are typically more pronounced in their change of direction than bends.

In small quantities, metal sheet can be folded by gripping it in a bending bar and hitting it with a mallet. For example, a metal tray may include folded tabs and folded over edges (see Figure 2.3.2). In larger quantities, presses powered by hydraulic rams are used to apply pressure along the fold line, to push the metal sheet into a former. The former helps to ensure accuracy and consistency.

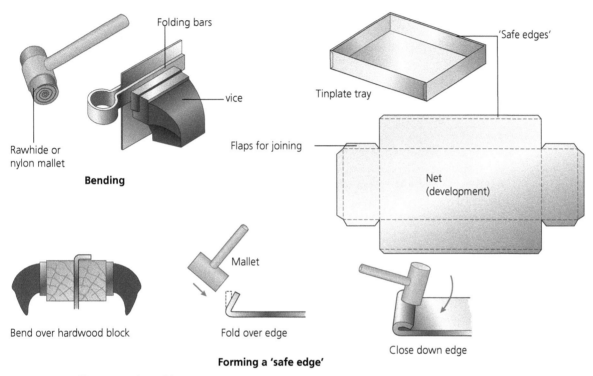

Figure 2.3.2 **Sheet metal working**

Press forming

Press forming is used in industry to make 3D shapes from metal sheet. The shapes can range from simple domes or bowls to more complex shapes like car body panels. The complete shape is made in a single operation from a sheet of ductile metal.

The process involves placing a sheet of material between either two moulds or a ram and die (see Figure 2.3.4). The ram and die or moulds fit together with a thin gap between them in the shape required. A press then applies pressure to permanently deform the metal sheet. For a simple profile where only a few products are required, a fly press can be operated by hand. For large parts, especially where many products are needed, a press powered by a hydraulic ram will be used. The pressure applied by this type of press can be many hundreds of tons. The excess metal around the shape may then be removed.

The moulds are typically made from high-carbon steel. As it is very hard, it resists wear from repeated use. However, this means that it is also difficult to machine, so moulds can be expensive. The cost of the mould will normally be divided between the number of parts that the mould is used to make.

Figure 2.3.3 **Worker operating a folding machine**

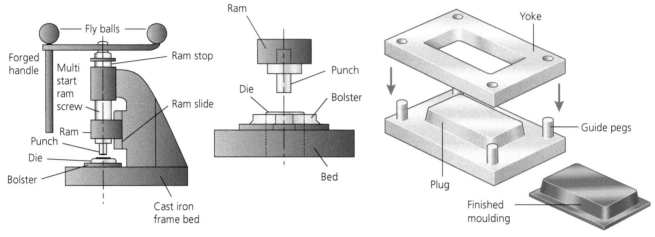

Figure 2.3.4 **Fly press**

Figure 2.3.5 **Press moulding**

Press moulding

Polymer sheets can be formed into shapes using a similar process, called press moulding. A sheet of thermoplastic polymer is first heated until flexible. It is then placed between two parts of a mould called a 'plug and yoke'. The plug is similar to a male mould and the yoke is like a female mould. The two parts of the mould are pressed together and the plastic is then allowed to cool until it hardens in the new shape. It is then taken out of the mould and any excess material removed.

KEY WORDS

Folding: bending a material over on itself.

Press forming: a forming operation used to change the shape of a sheet material using pressure from a press.

Punching: a process where a hole is made in a sheet and the material pushed out is scrap.

Stamping: a process where a part is pushed out of a sheet making a hole and the remaining material that was around it is scrap.

Mould: a hollow container used to give shape to a piece of material.

Punching and stamping

Punching and **stamping** can be used to cut shapes in metal sheet. In both techniques, pressure is applied to push a tool through the metal sheet.

In punching, a hole is made and the material pushed out is scrap. This is also known as piercing.

In stamping, the shape pushed out of the sheet is the desired part and the metal around it is scrap. This is also known as blanking.

Punching and stamping are often carried out simultaneously with press forming. They can be carried out using similar presses with either a ram and die or a tool integrated into a mould.

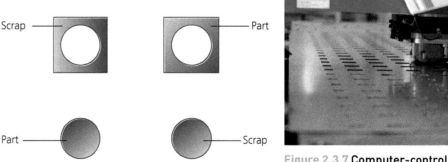

Figure 2.3.6 **Stamping and Punching**

Figure 2.3.7 **Computer-controlled machine punching holes in a metal sheet**

ACTIVITY

Many engineered products you use every day are made from sheet metal or pipe. Identify a different product that uses each of the following processes: bending, folding, press forming, punching and stamping. Try to come up with your own examples which haven't been used in this book.

MATHEMATICAL UNDERSTANDING

E10 Calculating pressure

Question

A press is being used to stamp components from a metal sheet. The area over which the tool contacts the sheet is 75 mm². The force applied by the press is 30 MN.

Calculate the pressure that the tool applies to the metal sheet.

Solution

The pressure $P = \dfrac{F}{A} = \dfrac{3 \times 10^7}{75} = 400 \ kN\,mm^{-2}$

Composite lay up

One of the most common composite materials is glass-reinforced plastic (GRP). This is made from glass fibres (which provide the reinforcement), surrounded by a matrix of a polymer. Composite materials of this type can sometimes be identified by their appearance, as they are typically smooth on one side and rough on the other. This is a result of the manufacturing process used to make them. They are made by hand using a **mould**. The smooth side is in contact with the mould. If using a male mould, this is the inside surface, and if using a female mould, this is the outside surface.

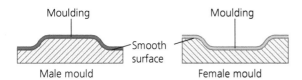

Figure 2.3.8 **Male and female moulds for composite lay up**

The process of making a product from GRP is as follows (see Figure 2.3.9).
- The mould is polished and coated with a release agent. It is then covered in a gel-coat resin. This gives a good finish on the smooth side and may also be used to add colour.
- Layers of the reinforcement material (glass fibre) are placed in the mould. Care is taken to ensure that the reinforcement is pushed in to any corners.
- The reinforcement is then soaked in resin to form the composite matrix. This may be painted or stippled (dabbed on with a brush) on to it.
- More layers of reinforcement are added and soaked with resin until the required thickness is achieved.
- Where high performance is required, the entire mould and part might be encapsulated in plastic sheets and a vacuum applied to suck out the air. This ensures that resin has been pulled into all the spaces in the glass fibre to reduce the risk of local areas of weakness.
- The part is cured, which causes the resin to become hard. This may involve either allowing it to stand for twenty-four hours, heating it gently in an oven, or putting it in a machine called an autoclave, which applies both heat and pressure.

Composite lay up

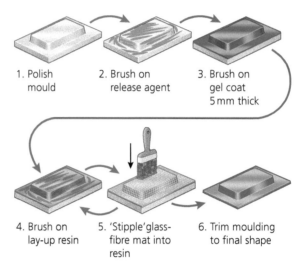

1. Polish mould

2. Brush on release agent

3. Brush on gel coat 5 mm thick

4. Brush on lay-up resin

5. 'Stipple' glass-fibre mat into resin

6. Trim moulding to final shape

Figure 2.3.9 **GRP moulding**

KEY POINTS

- Thin sections of metal can be bent when cold. Both metals and polymers are easier to bend when heated. Bending jigs are used when an accurate bend or particular shape is needed.
- Folding can be carried out using a bending bar and mallet or a hydraulic press and former.
- Press forming is used to make 3D shapes from metal sheet using a ram and die, or a press and mould.
- Punching and stamping are used to cut shapes in metal sheet.
- The lay-up method uses a mould to shape a composite product. The reinforcement is placed into the mould and the matrix applied so that it fills the gaps around it.

Check your knowledge and understanding

1 Give two reasons for using a jig when bending a batch of parts.
2 Suggest a suitable material to make a mould for press forming. Explain your choice.
3 Explain how the use of a male mould, rather than a female mould, could affect a composite product made using the lay-up method.

2.4 Casting and moulding

What will I learn?

By the end of this section you should have developed knowledge and understanding of:

→ how sand casting and pressure die casting are used to make products from metal
→ the differences between the moulds used for sand casting and pressure die casting
→ how injection moulding is used to make products from polymer.

Figure 2.4.1 **Metal casting**

Casting and moulding are shaping processes. Melted metal or plastic is poured or forced into a mould and then solidifies to take on the shape of the hollow area within the mould. The hollow area is called the cavity.

It is normally much more cost effective to make a complicated 3D shape by casting or moulding than by machining it from a larger block of material; casting or moulding reduce the labour time needed and the waste produced by the machining. It is also much more cost effective than joining together many smaller pieces of material to make the shape also due to the labour time that this would take.

Sand casting

Sand casting is used to make metal parts. It gets its name from using a mould that is made from bonded sand. This is sand mixed with chemicals that help it to stick together. Sand casting is commonly used for one-off parts or where small quantities are needed. If large quantities of parts are needed, it would take a lot of time to make many sand moulds and, unless the moulds are lined up perfectly every time, there is a risk of small variations between products.

The sand mould is made using a reusable pattern. This is a wooden version of the product that is to be cast. The pattern should have no sharp corners which can be difficult to cast. It is normally split into two parts which are used separately to form the top and bottom of the mould.

Stage 1

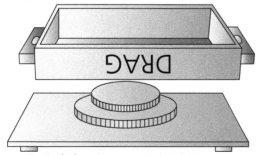

One half of a split pattern is placed onto a moulding board and the bottom half of the moulding box (the drag) is placed upside down over the pattern.

Stage 2

The pattern is cut through the centre and fitted with location dowels.

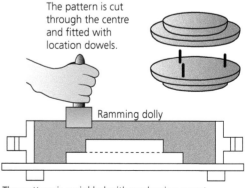

Ramming dolly

The pattern is sprinkled with a releasing agent (parting powder). Sand (Petrobond) is then added to the moulding box and rammed around the pattern.

Stage 3

The top is levelled off (strickled) and the whole assembly turned over.

Stage 4

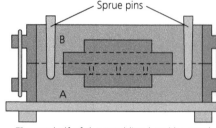

The top half of the moulding box (the cope) is then placed on top of the drag. The top half of the pattern is then fitted to the bottom half. Sprue pins are fitted, sprinkled with parting powder, and once again Petrobond sand is rammed around the pattern.

Stage 5

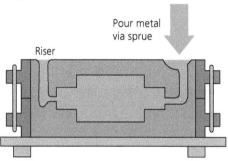

The pins are then removed, the top half of the mould taken off and the patterns are removed. Channels are then cut to connect the runner and riser to the mould cavity, and then the mould is reassembled and is ready for casting.

Figure 2.4.2 **Sand casting process**

Molten metal is poured into the mould through the sprue and allowed to solidify. Each sand mould is used just once, as the sand is knocked or shaken off the casting. However, the sand can be reused to make other moulds. The sprue, riser and any runners can be cut off the cast part. If necessary, the part may also be machined to improve the surface finish.

Pressure die casting

Pressure die casting is mainly used to make parts from non-ferrous metal. It uses a special type of mould called a 'die', normally made from high-carbon steel. The die typically has two halves and must have a melting point higher than the metal being cast. It is normally machined to a smooth finish and may include channels to allow coolant to flow through it. As the metal used to make the die is very hard (to resist the wear of repeated use), it is difficult to machine and often very expensive to make. This means that pressure die casting is typically used only when many parts are required so that the cost of the equipment and die can be divided between all the products being made.

During the casting process (see Figure 2.4.4), the parts of the die are brought together. This pressure forces the liquid metal into the cavity. Once the metal has solidified, the die is taken apart and the cast part removed. The use of pressure means that the castings are typically more accurate than sand casting, with more detailed features and a better surface finish.

Figure 2.4.3 **Steel mould used in die casting**

KEY WORD

Pressure die casting: a shaping process where molten metal is forced into a reusable metal mould.

2.4 Casting and moulding

Permanent mould
Molten metal is forced into a water-cooled metal mould (die) through a system of sprues and runners. The metal solidifies rapidly and the casting is removed with its sprues and runners.

SHAPE
3D solid
Used for complex shapes and thin sections. Cores must be simple and retractable.

MATERIALS
Light alloys
High fluidity requirement means low melting temperature eutectics usually used. Hot chamber method restricted to very low melting temperature alloys (e.g.Mg)

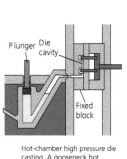

Hot-chamber high pressure die casting. A gooseneck hot chamber is submerged in a pot of molten metal. Metal is injected directly from the pot via the gooseneck.

CYCLE TIME	QUALITY	FLEXIBILITY	MATERIALS UTILIZATION	OPERATING COST
Solidification time is typically<1 s so cycle is controlled by time taken to fill mould and remove casting.	Good surface texture but turbulent mould filling produces high degree of internal porosity.	Tooling dedicated so limited by machine setting up time.	Near net shape process but some scrap; in sprues, runners and flash which can be directly recycled.	High, since machine and moulds are expensive.

Figure 2.4.4 **Pressure die casting**

STRETCH AND CHALLENGE

Identify a metal engineered component that has been made using a casting process. Can you identify all the manufacturing operations that would need to be carried out if this component had to be manufactured from a solid block using conventional cutting processes instead?

Injection moulding

Injection moulding is a similar process to pressure die casting but it is used for parts made from polymer. It is a very versatile process, used to produce a wide range of items such as electrical plugs, model construction kits, toys and plastic boxes.

The steps involved in producing an injection moulded part are as follows.
● Polymer granules are loaded into a feed hopper. This feeds material into the barrel of the injection moulding machine.
● A rotating screw thread pushes the polymer along and it is melted by heaters.
● At the end of the heating area, the liquid polymer is compressed by the cone shape as the diameter of the barrel reduces. This increases the pressure on the polymer and it is forced into the mould.
● The mould is cooled and the component is ejected.
● The mould is moved back into position to make the next part.

The equipment and moulds are normally very expensive. However, for small parts, the process of injection, cooling, ejection and repositioning the mould can be completed in only a few seconds meaning that hundreds or thousands of parts can be produced each hour. This allows the high cost to be divided between all the parts made.

> **KEY WORD**
>
> **Injection moulding**: a shaping process for polymers, where the polymer is forced into a reusable metal mould.

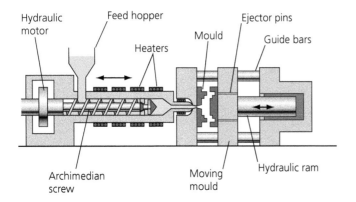

Figure 2.4.5 **Injection moulding machine**

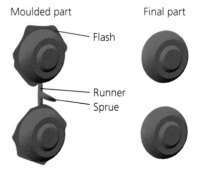

Figure 2.4.6 **Flash on an injection moulded part**

Injection moulded products can sometimes be identified visually. They may have a visible 'sprue point' where the plastic was injected to make them, and the sprue subsequently cut off. They may also have a 'parting line' which is the line of separation on the part where the two halves of the mould meet. If the sides of the mould did not fit together perfectly then there may also be excess polymer, called 'flash'.

ACTIVITY

Find a variety of injection moulded products in your classroom or workshop. Can you identify the parting line or sprue point on any of these?

KEY POINTS
● Sand casting and pressure die casting are used to make 3D metal parts.
● Sand casting uses a mould made from sand, which is destroyed after use.
● Pressure die casting uses a reusable mould made from metal. Due to the cost of the equipment it is typically only used when a large quantity of parts is to be made.
● Injection moulding is similar to pressure die casting, but for parts made from polymer.

Check your knowledge and understanding

1 Explain the benefits of using sand casting rather than conventional machining to make complex 3D products.
2 Explain what is meant by the term 'pattern' when making a mould for sand casting.
3 Describe in detail the process of injection moulding.

Joining and assembly

What will I learn?

By the end of this section you should have developed knowledge and understanding of:

→ the types of threaded fastenings used to attach materials together and the reasons why they may be selected for an application

→ how riveting is used to join sheets of material together

→ the differences between soft and hard soldering, brazing and welding and how these processes are used to join materials together.

Joining and assembly involves attaching or putting parts together. The joining process used is normally specified in the design as it can have a significant effect on the strength of the product. If it may be necessary to take the assembly apart again, a 'temporary' joining process may be used, such as nuts and bolts or screws. For increased strength, a 'permanent' joining process may be used, such as brazing or welding. Separating parts that have been joined using a permanent process would normally result in damaging one or more of the parts.

Threaded fastenings

KEY WORD

Threaded fastenings: products such as screws, nuts and bolts that can be used to make a temporary joint.

Threaded fastenings include nuts, bolts and screws. These are available in a wide range of materials, including steel, brass and thermoplastic polymers. They can be used to make joints in most materials, and can incorporate a variety of materials within one product (e.g. joining polymers to metals).

When assembling products in small quantities, threaded fasteners can be attached by hand using screwdrivers or spanners, depending upon the head-type of the fastener. When assembling larger quantities of products, powered tools such as pneumatic screwdrivers might be used. These work faster and reduce how quickly the assembler gets tired.

An advantage of threaded fasteners is that they can be later removed, meaning a product can be taken apart, for example, to open a panel to allow maintenance, repair or the replacement of batteries, or if a product needs to be disassembled to recycle the different materials used to make it. A limitation of this type of temporary fixing is that it may become loose when in service.

Figure 2.5.1
Hexagonal bolt head

Wing nut   Hexagonal nut   Square nut   Locking nut   Castle nut and split pin   Nylon (fibre lock) nut

Figure 2.5.2 **Types of nut**

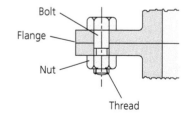

Bolt

Flange

Nut

Thread

Figure 2.5.3 **Flange secured with a nut and bolt**

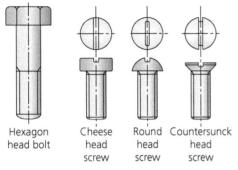

Hexagon head bolt   Cheese head screw   Round head screw   Countersunck head screw

Figure 2.5.4 Standard representations of machine screws

Figure 2.5.5 Typical machine screw

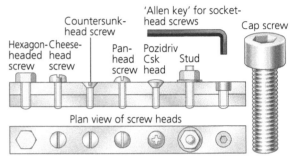

Figure 2.5.6 Types of machine screw

Rivets

Rivets are used to hold sheets of material together, for example, attaching overlapping metal plates to form the hull of a ship or attaching the skin to an aircraft. Most rivets are used to join metal sheets, although they can also be used to join polymers, leather or different materials.

A hole is drilled in overlapping sheets of material. The rivet is inserted through the hole. The ends are then hammered over to mechanically hold the sheets in place.

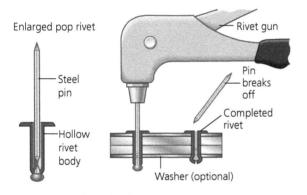

Figure 2.5.8 Pop riveting

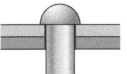 

Figure 2.5.7 Simple riveted joint

Pop rivets

A disadvantage of rivets is that access is needed to both sides of the joint. This can be overcome by using pop rivets (see Figure 2.5.8). A pop rivet has two parts: the pin and the rivet. The pop rivet is placed in the hole through the sheets to be joined. The rivet gun then pulls the pin through the rivet. This deforms the end of the rivet so that it joins the sheets. The pin then breaks away, leaving a uniform shape on the visible surface of the joint. Pop riveting is used where the metal or polymer is thin and normally forms a weaker joint than solid riveting.

KEY WORD

Soldering: a joining process where metal parts are attached together using a filler wire which melts and runs between them, typically melted using a soldering iron.

Soldering

Soldering is a process in which two (or more) metal parts are joined together. It involves melting solder to form a joint between the pieces being joined. Solder is an alloy of another metal, which has a much lower melting point than the metals being joined. If the solder melts at temperatures below 450 °C, it is called soft soldering. If solder melts at temperatures above 450 °C, it is called hard soldering.

None of the metal parts being joined are melted. The solder flows into any spaces between the metals being joined and physically holds them in place. It is sometimes referred to as a 'filler metal' as it fills any spaces in the joint.

Soft soldering

Soft soldering is very commonly used in school workshops to attach components on to a printed circuit board (PCB). The solder used for electrical applications is a tin alloy containing small proportions of copper, silver and other metals, with a melting point typically between 180 and 220 °C. Previously, a tin-lead alloy was used, but the use of this has been prohibited due to concerns about the impact of lead on health. The solder wire used may not be solid but contain a core of flux. This is a chemical designed to prevent oxidation of the metals to ensure the electrical conductivity of the joint.

The legs of the electrical components are placed through the holes in the PCB. A soldering iron is then used to heat the area where the component leg sticks out from the copper track. Solder wire is pushed into the joint, melts and flows into the space between the metal parts.

Wave soldering

When manufacturing large quantities of PCBs, 'wave soldering' may be used. This is where the PCB and assembled components are positioned just above a bath of solder. A ripple across the molten solder is sufficient to allow it to flow into the joints around the components, completing all the joints in a single operation. Areas where the solder is not required are protected by a film of material that the solder cannot adhere to.

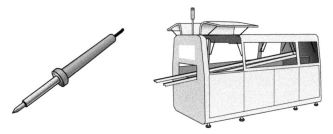

Figure 2.5.9 **Soldering iron and wave soldering machine**

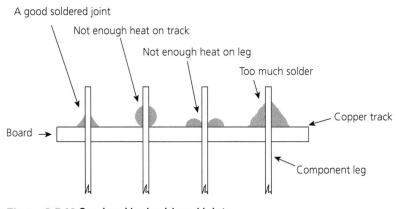

Figure 2.5.10 **Good and bad soldered joints**

Figure 2.5.11 **Soldering and electronic circuit**

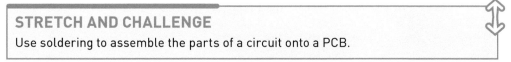

STRETCH AND CHALLENGE
Use soldering to assemble the parts of a circuit onto a PCB.

Hard soldering

Hard soldering is sometimes called 'silver soldering'. It can be used to join metal parts together using a filler wire and is often used when making gold or silver jewellery. Silver soldering differs from soft soldering in three main ways.

- The alloy used for the solder melts at a much higher temperature. It typically contains 30–80 per cent silver, with copper, zinc and other metals, and has a melting point of 620–740 °C.
- Due to the high melting point of the solder, the heat may be applied directly using a flame.
- A flux is typically applied directly to the area to be soldered. This prevents oxidation of the metals at their point of contact, helping to ensure that the solder can form a strong bond.

Brazing

Like hard soldering, **brazing** is used to join metal parts together using a filler wire. However, it is carried out at temperatures of 450–1200 °C, depending on the filler metal being used. Common types of filler metal include aluminium–silicon alloys, copper alloys (with silver, zinc [forming brass], or tin [forming bronze]), nickel alloys and alloys made from gold and silver.

The surface to be joined is first cleaned of oxides and rust, paint and grease. Similar to hard soldering, a flux is applied to prevent oxidation of the surface at the brazing temperature. The joint is then heated. If small quantities of parts are being brazed, this may be direct heating using the flame from a gas torch with the filler metal being fed in as the joint reaches the necessary temperature. If large quantities of parts are to be brazed, filler wire may be positioned in or by the joint, and the whole assembly heated in a brazing furnace. After the filler wire has melted and flowed into the joint, it is allowed to cool and harden.

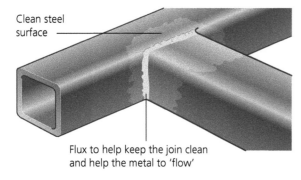

Clean steel surface

Flux to help keep the join clean and help the metal to 'flow'

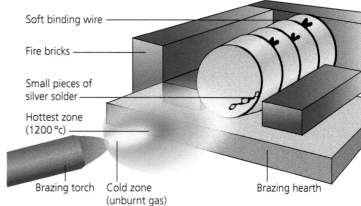

Soft binding wire

Fire bricks

Small pieces of silver solder

Hottest zone (1200 °c)

Brazing torch   Cold zone (unburnt gas)   Brazing hearth

Figure 2.5.12 Hard soldering process

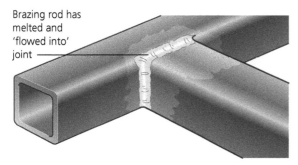

Brazing rod has melted and 'flowed into' joint

Figure 2.5.13 Brazing two pieces of square tube

Figure 2.5.14 Brazing copper pipes

Welding

Welding is used to join two parts made from similar metals together. As with brazing, the joint must be thoroughly cleaned before the process and a filler metal is used where there is a gap between the parts being joined.

The principal difference between welding and brazing is that, in welding, the edges of the parts being joined are melted. To achieve this, welding is carried out at a higher temperature; the heat source used may have a temperature of over 3000 °C. This may be obtained by:

- burning acetylene gas with pure oxygen, known as oxyacetylene welding
- creating a spark between a non-consumable electrode and the workpiece, with filler wire added separately, known as tungsten inert gas (TIG) welding
- creating a spark between a consumable electrode and the workpiece, where the electrode melts to form the filler wire. This is known as metal inert gas (MIG), metal active gas (MAG) or manual metal arc (MMA) welding.

Arc welding

When an electric spark is used, it is called an arc. The different processes that use this (TIG, MIG/MAG and MMA) are called 'arc welding' processes.

The heat source continually moves along the joint, melting a small area directly where the heat is applied (called the 'weld pool'). As the heat source moves on, the area behind it quickly cools and solidifies. Due to the high temperatures, care must be taken to prevent oxidation of the metals as this will lead to a weak joint. In TIG, MIG and MAG welding, a shielding gas flows over the arc and joint to separate it from the atmosphere. In MMA welding, flux from the filler rod covers the weld, and must be chipped off and cleaned away after the joint has cooled.

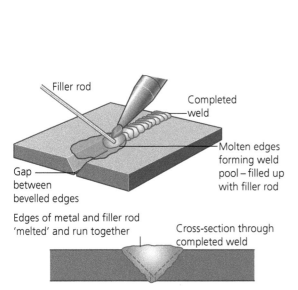

Filler rod
Completed weld
Gap between bevelled edges
Molten edges forming weld pool – filled up with filler rod

Edges of metal and filler rod 'melted' and run together
Cross-section through completed weld

Figure 2.5.15 **Oxyacetylene welding**

Figure 2.5.16 **MIG welding set**

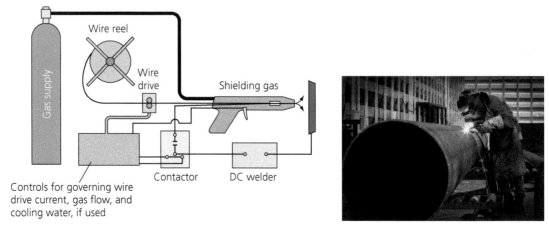

Figure 2.5.17 **Schematic of MIG welding equipment**

Figure 2.5.18 **Welding a pipe**

Although arc welding processes are the most common in engineering, there are welding processes that use other sources of energy. These include spot welding, which uses heating caused by electrical resistance in a joint (see Figure 2.5.19), laser welding, and friction welding, which uses heat caused by friction.

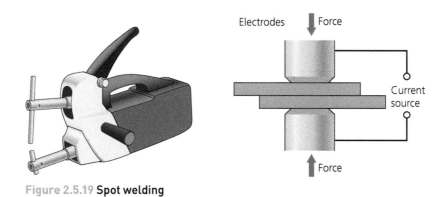

Figure 2.5.19 **Spot welding**

ACTIVITY

Create a table to show which joining processes can be used with different materials. Choose five specific materials of different types. Types could include: ferrous metal, non-ferrous metal, polymer, composite, timber or ceramic etc. Each is written in both the rows and columns, as shown in the example table. In each box, write in all the joining processes that could be used to join the two materials, in order of strongest to weakest joint. An example has been done for you.

Material:	Low-carbon steel	Aluminium	ABS	GRP	Structural timber
Low-carbon steel	Welding, brazing, hard soldering, rivets, threaded fastenings				
Aluminium					
ABS					
GRP					
Structural timber					

KEY POINTS

- Threaded fastenings, such as screws or nuts and bolts, are used to make temporary joints in a wide range of materials.
- Rivets and pop rivets can be used to join sheets of material together. Rivets require access to both sides of a joint; pop rivets need access only from one side.
- Soldering and brazing use a filler wire that is melted and runs between the metal parts to form a joint. The metal parts being joined are not melted. Soldering is carried out at a lower temperature than brazing.
- Welding involves melting the edges of the metal parts being joined to form the joint, with filler wire added if necessary. The temperature reached by the joint is much higher than that for brazing.

Check your knowledge and understanding

1 Name three processes that could be used to join a sheet of metal to a sheet of polymer.
2 Name two tools used to attach threaded fasteners to products.
3 Describe how rivets are used to attach sheets of material together.
4 Describe how soldering is used to attach a component to a printed circuit board.
5 Explain the differences between joints produced using brazing and welding.

2.6 Heat and chemical treatment

What will I learn?

By the end of this section you should have developed knowledge and understanding of:

→ how normalising, annealing, quenching and hardening are carried out
→ how these processes affect the properties of the metals that they are used with.

Figure 2.6.1 Heat treatment furnace

Heat treatment can be used to change the properties of metals, either by increasing the grain size within the material or changing how the atoms within the grains are arranged. The reasons for this were explained in Section 1.

Normalising

Normalising is carried out on steel that has been work hardened. For example, it may have been formed by applying great force, such as using the forging process. It involves heating the steel just above its upper critical point then allowing it to cool naturally in air. Normalising provides sufficient time for the atoms to rearrange within the existing grains, relieving internal stresses. This results in steel that is tough with some ductility. However, unlike annealing it does not allow the grains to grow, so the steel does not soften a great deal.

Annealing

Annealing involves heating the metal to a suitable temperature and holding it there for a given time. This allows the grains within the metal to grow making the metal softer and easier to work. This also relieves the internal stresses and allows the dislocations to relocate to the grain boundaries. It is often used to soften metal that has been work hardened, or to make metals easier to bend into complicated shapes.

Annealing is used for both ferrous and non-ferrous metals.

- Steel is annealed by heating it to just below the lower critical point. It is then allowed to 'soak' at that temperature for a period of time. The duration depends upon the size of the part. The metal is then slowly cooled, typically by leaving it in the furnace after it has been switched off.
- Aluminium alloys are often annealed after working. Care must be taken to control the annealing temperature of 350–400 °C; if it is overheated it can approach its melting point of 660 °C.
- Copper is easy to anneal. It is simply heated until it is a dull red colour, then either cooled in water or left to cool in air. However, black scale (corroded copper) may form on the surface due to corrosion; this needs to be cleaned off by putting the copper into a bath of dilute acid, which is known as 'pickling'.
- Brass can also be annealed by heating to a dull red colour. However, it is brittle at this temperature which is known as 'hot shortness'. This means that it must be cooled slowly. Like copper, it needs pickling to remove black scale.

Hardening and quenching

High-carbon steel, containing 0.8–1.4 per cent carbon, can be hardened by heat treatment; low-carbon steels cannot be hardened in this way.

The steel is first heated to a temperature just above the lower critical point. It is then allowed time to soak at that temperature giving the atoms within the grains time to rearrange and to form 'austenite'. The steel is then put through the process of **quenching** which involves cooling it rapidly by immersing it in oil or brine (salt water). This does not give the atoms time to rearrange again, so the structure becomes 'martensite'.

Tempering

As the martensitic structure is so hard and brittle, it must be put through the process of **tempering**. This involves heating it to a temperature of 230–300 °C, then quenching it again in oil or brine. Tempering removes some of the hardness and internal stresses, and reduces the brittleness so that the metal part is less likely to shatter when used. The precise tempering temperature used will be selected based on the hardness and toughness required depending upon what the steel is going to be used for.

KEY WORDS

Hardening: a heat treatment that increases the hardness and strength of a metal due to a change in the arrangement of the atoms within it.

Quenching: the rapid cooling of a hot metal by immersing it in a liquid, often oil or brine.

Tempering: a heat treatment to remove some of the brittleness in a hardened steel, at the cost of some hardness.

Colour	Temp. °C	Hardness	Typical Uses
Light Straw	230	Hardest	Lathe Tools, Scrapers
Dark Straw	245		Drills, Taps and Dies, Punches
Orange/Brown	260		Hammer heads, Plane irons
Light Purple	270		Scissors, Knives
Dark Purple	280		Saws, Chisels, Axes
Blue	300	Toughest	Springs, Spinners, Vice jaws

Figure 2.6.1 **Tempering colours and temperatures**

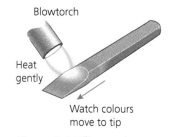

Figure 2.6.2 **Tempering a cold chisel**

Blowtorch

Heat gently

Watch colours move to tip

ACTIVITY

Hand tools such as saw blades, screwdrivers and hammers are often heat treated. Make a table similar to the one below listing the metal tools in your school workshop. For each, state what form of heat treatment might have been used and provide a reason for your choice.

Tool	Heat treatment	Reason
Coping saw blade		
Screwdriver		
Chisel		
Hammer		

KEY POINTS

- Normalising involves heating steel then allowing it to cool in air. It results in steel that is tough with some ductility.
- Annealing involves heating a metal and allowing it to hold at a temperature where grain growth occurs. It makes metal softer and easier to work. Annealing can be used with steel, aluminium alloys, copper and brass.
- Hardening steel involves heating and quenching. This results in a hard but brittle material. Quenched steel is normally tempered, which reduces some of the brittleness at the expense of some of the hardness.
- Some aluminium alloys can be precipitation hardened, where elements dissolved in the alloy can increase the strength.

Check your knowledge and understanding

1 State what is meant by normalising.
2 Name three metals or alloys that can be annealed.
3 State what is meant by 'quenching'.

2.7

Surface finishing

What will I learn?

By the end of this section you should have developed knowledge and understanding of:

→ the reasons why a surface finish may be applied to a material
→ how paints and polymers are applied to metal products
→ how electroplating allows coatings to be formed on metal products
→ how steel is galvanised using zinc
→ the methods and reasons for polishing materials.

Surface finishing involves changing the surface of a part or product in a useful way, such as:

● improving the material's resistance to corrosion
● increasing the material's ability to resist scratches and damage
● improving the material's appearance, making it more attractive to the end user.

For some applications, this can mean that a lower cost material with a coating can be used rather than making the whole product from a more expensive material with better properties. It can also mean that the properties of different metals can be combined, such as the strength and toughness of steel with the corrosion resistance of zinc.

When using most surface finishing processes, it is important that the part to be coated is thoroughly cleaned first, to remove oxides (such as rust), dirt and grease. If not, these may result in areas that are uncoated or where the coating does not attach to the part and will flake off.

Painting

Painting is probably the most common surface finishing process used. It can increase corrosion resistance and improve the visual appearance of metal surfaces. Compared to other coating processes, it is relatively quick to carry out and low cost, particularly when finishing small numbers of parts.

Paints consist of three components:
● pigment: to provide the colour
● vehicle: makes the paint adhere to the surface and forms a film as the paint dries
● solvent: which evaporates as the paint dries and controls how easy the paint is to apply.

When only a small number of parts need to be finished, paints can be applied manually by brushing. When larger quantities of products are needed, paint can be sprayed, either manually or using robot arms, or applied by dipping.

Figure 2.7.1 **Spray painting a car body part**

Dip coating

Dip coating is used to apply polymer coatings such as PVC, nylon or polyethylene to metal parts. It provides a thick, wear-resistant finish that keeps water and air from the surface of the metal, preventing corrosion. The process of dip coating involves the following steps.

● The metal part is cleaned thoroughly to remove any oxide and rust, dirt and grease.
● It is then heated in an oven, typically to 250–400 °C.

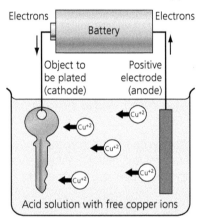

- When hot enough, it is dipped in a fluidised bed of polymer particles. A fluidised bed is a tank of polymer powder through which air is being blown from the bottom making the polymer powder rise and bubble.
- The part is held in the tank as polymer particles stick to its surface and melt, until either the required thickness is achieved or the metal has cooled so no more polymer is attaching itself. The part is then removed from the tank.
- The coated metal part is then reheated briefly to ensure that all the attached polymer powder has melted. It is then allowed to cool.
- The coating should be even and shiny. If there are any pin holes or other defects, or if a greater thickness of plastic is needed, it can be reheated and given another coat.

Electroplating

Electroplating uses electricity and a chemical solution to create a coating on a metal part.

Electrolysis

The most commonly used form of electroplating uses a process called '**electrolysis**'. This coats a thin layer of metal onto the surface of a part. The coating is typically nickel, zinc, copper or tin.

The part to be coated is put into a bath or tank containing a solution of chemical salts. A piece of the metal that will be used for the coating is put in the tank and attached to the positive side of an electricity supply. This is called the anode. The part to be coated is attached to the negative side of the electricity supply. This creates a circuit, as electricity can flow through the chemical solution. As the current flows, the anode gradually dissolves away and a thin layer of the metal slowly builds up on the part. The amount of metal deposited depends upon the strength of the current and the concentration of the solution.

Some forms of electroplating do not dissolve the anode. In these, the metal for the coating is extracted from the chemical solution. It is important that the concentration of metal salts in the solution is increased or topped up, otherwise the process becomes less effective. Chromium plating is produced in this way. Chrome-plated parts are normally nickel plated first to improve corrosion resistance, before the chromium plating is used. The chrome provides an attractive shiny finish.

Figure 2.7.2 Electroplating

Anodising

Anodising is an electroplating process used on aluminium or titanium when electrolysis is carried out in an acid solution. The coating is a relatively thick oxide layer. It provides a corrosion resistant, durable finish and colour can be added to tint the coating, making it more attractive.

Anodising is used to make the dielectric films used in electrolytic capacitors, dental implants and coloured jewellery.

Galvanising

Galvanising is an industrial process that involves dipping steel in a bath of molten zinc. The liquid zinc sticks to the steel forming the coating. The zinc provides good corrosion resistance, although the coated part is not visually attractive. Galvanised steel is used to make metal dustbins and buckets. It is often painted to improve its appearance before it is used in commercial products, such as car body panels.

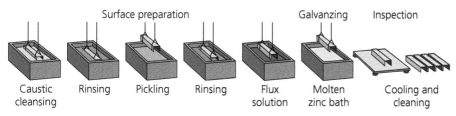

Surface preparation — Galvanzing — Inspection

Caustic cleansing | Rinsing | Pickling | Rinsing | Flux solution | Molten zinc bath | Cooling and cleaning

Figure 2.7.3 **A typical galvanising line**

Polishing

Polishing is a physical process that gives a material a shinier appearance. It also makes the surface smoother, reducing friction.

Polishing involves removing a tiny amount of material of the surface of the part. This can be carried out using a buffing wheel or a non-abrasive cloth. It is used for both metals and, when needed, the edges of polymer products. Polishing is the only surface-finishing process used for polymers in the engineering industry. Polymers are otherwise thought to be 'self-finishing'.

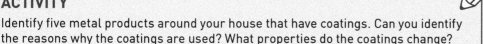

ACTIVITY

Identify five metal products around your house that have coatings. Can you identify the reasons why the coatings are used? What properties do the coatings change?

KEY POINTS
- Painting is a relatively quick and low-cost way to increase the corrosion resistance and improve the appearance of a bare metal part.
- Dip coating is used to apply a polymer coating to a metal part. The metal part is heated and placed in a fluidised bed of the polymer.
- Electroplating involves placing a metal part in a chemical bath and using electricity to build up a coating.
- Galvanising applies a coating to steel by dipping it in molten zinc.
- Polishing involves removing a tiny amount of material of the surface of the part making it smoother and shinier.

Check your knowledge and understanding

1. State two methods used to apply paint to a product.
2. Describe the process of dip coating.
3. State four metals that can be coated onto steel by electrolysis.
4. Name three steel products that are galvanised.
5. Give two reasons for polishing a metal product

PRACTICE QUESTIONS: engineering manufacturing processes

1 Explain one advantage and one disadvantage of using rapid prototyping to make a part.

2 Describe how a part is manufactured using fused deposition modelling.

3 Explain how a metal sheet is cut by a guillotine.

4 Describe what is meant by 'taper turning'.

5 State two machines that can be used to make a hole down the middle of a solid metal cylinder.

6 Describe the process of making a printed circuit board.

7 Explain the difference between forming and shaping.

8 Name two tools or machines used to fold metal sheet.

9 Explain the difference between punching and stamping a metal sheet.

10 Using notes and/or sketches, describe how a composite material is made by the lay-up method.

11 Explain the differences between the moulds used for sand casting and pressure die casting.

12 Name two types of threaded fastenings.

13 Describe the process of pop riveting.

14 Name three welding processes.

15 Explain how annealing changes the properties of an aluminium alloy.

16 Give three reasons for using a surface finish on a metal part.

17 Name two methods that can be used to coat a low-carbon steel part with zinc.

3 Systems

Engineered systems are fundamental to how we conduct our daily lives. From the electronic security systems that protect our homes, to the pneumatic systems used to produce many of the food products that we eat, they play an important role in the world that we live in today. Because of this, it is vital that engineers can analyse, understand and affect how systems work. Developing new and more advanced systems allows us to solve more problems that affect people, and therefore further improve their quality of life.

Systems can be mechanical, electrical, electronic, structural, fluid-based or a combination of two or more of these different types. This section will enable you to understand the basics of systems so that you can apply and make use of them in your own engineering projects.

This section includes the following topics:

3.1 Describing systems

3.2 Mechanical systems

3.3 Electrical systems

3.4 Electronic systems

3.5 Structural systems

3.6 Pneumatic systems

At the end of this section you will find practice questions relating to systems.

3.1 | Describing systems

What will I learn?

By the end of this section you should have developed knowledge and understanding of:

→ how block diagrams are used to represent systems
→ how schematics are used to communicate circuit designs
→ the main symbols used to construct flowcharts.

This section describes the three main ways of communicating and describing a system:

● system block diagrams
● circuit schematics
● flowcharts.

This section will introduce the purpose of each of these, how they are drawn and the typical uses of each. It also investigates the parts that make up a system, namely their inputs, processes and outputs. By the end of this section you should understand what a system is and how different systems can be effectively represented and described.

A range of examples of these are given throughout this section. All systems presented in this book are described using one or more of these methods.

System block diagrams

A **system**, as a whole, is made up of a collection of components, parts or **sub-systems**, that work together to achieve an outcome or function. Systems are described in terms of their:

● **inputs**: take a signal from the real world and convert it into a signal that the process block will understand, for example, converting light, sound or movement into an electrical current or voltage
● **processes**: change the signal in some way, for example, by increasing or decreasing its size or latching it on for a period of time. They are often thought of as the 'brain' of the system
● **outputs**: turn the signal back into a real-world signal, such as motion or heat.

Systems are represented and communicated using block diagrams. These show how the system will work on a 'top-down' level. This means starting with a basic overview of the whole system. The specifics of each sub-system, such as a full list of all individual components needed, are then considered later. This is known as the '**systems approach**'.

How to describe systems

On a **system block diagram**, the blocks represent the functions or sub-systems. The arrows represent the signals that are sent from and to each block. A common mistake when drawing block diagrams is to forget to include these signals, or assuming that the arrows for these are part of each function.

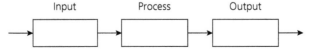

Figure 3.1.1 **System block diagram**

An example of a simple electronic system is an outside security light that automatically lights up for a timed period when it goes dark.

Figure 3.1.2 **Block diagram for an outdoor security light**

In the example above, the light sensor would detect when it is dark and send this information to the timer sub-system. This would then turn on the LED lamp, producing light for a set period of time before switching it off.

Systems with multiple inputs and outputs

A common mistake when designing systems is assuming they can only have one single input process and a single output sub-system. But a system can actually incorporate several inputs and several outputs together. For example, a robot arm would make use of several sensors and different types of motors to operate correctly.

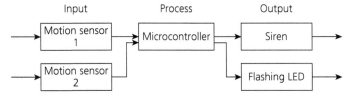

Figure 3.1.3 **Block diagram for a burglar alarm**

A burglar alarm is another good example. In this system, there could be multiple motion sensors placed around the building to sense when burglars have entered. There could also be a combination of output devices to produce light and sound when the alarm is activated. A microcontroller could be programmed as the main process control device in the system. The designer could easily add more sensors or even a keypad to the design by extending the block diagram further.

ACTIVITY

Draw a block system diagram for a robot buggy. The buggy must be able to detect when it is about to come into contact with objects in its path. It should then take evasive action to avoid hitting them.

ACTIVITY

Draw a schematic for a circuit that you have studied. Label all the components used.

Schematic drawings

Where block diagrams show a top-down representation of a system, **schematics** show the individual components required and how they are connected together. They use **circuit symbols** to represent the individual components used. Circuit symbols are based on agreed standards, which reduces confusion when reading them. They also take up much less space than diagrams drawn pictorially. A full list of symbols used in engineering schematics appropriate to this course is given in the Engineering symbols section of this book (p. 218).

An example of a simple electronic schematic (also known as a circuit diagram) for a simple transistor based circuit is shown in Figure 3.1.4. This circuit would turn on the buzzer when the light level falls to a certain level. The schematic clearly shows how all the components connect to complete the full circuit. This would become very difficult to understand if it was drawn as a picture.

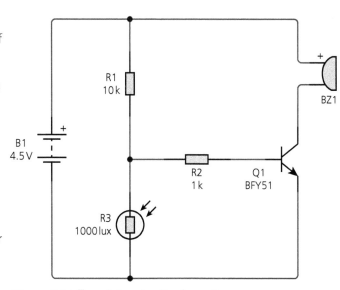

Figure 3.1.4 **Transistor circuit schematic**

When drawing schematics, it is good practice to label each individual component. The label should indicate the type of component and an identifying number if there is more than one of the same type in the circuit. For example, resistors used in an electronic circuit would be labelled as R1, R2, R3 etc. The value of components should also be shown where appropriate.

Flowcharts

Flowcharts are used to show the order in which a set of events is carried out. For example, they can be used to show how a set of manufacturing processes is carried out or how quality control procedures are applied to it. Another common use of flowcharts is when programming microcontrollers or other programmable components. This particular use is covered further in Section 3.4 Electronic Systems.

Flowcharts are often confused with system block diagrams but they are very different in terms of their purpose, what they represent and how they are drawn.

Drawing flowcharts

Flowcharts are drawn using symbols as listed by the British Standards Institute. Some of the most common symbols and their functions are shown in the Table 3.1.1.

Symbol	Name of Symbol	Meaning
	Terminator	Used to indicate the beginning and end of the series of events shown in the flowchart.
	Input/Output	Used to show information, data or other items that are either received (input) or given out (output).
	Process	Used to show when an action or instruction is to be carried out.
	Decision	Used to show where a choice must be made or a question answered. This must be in the form of a yes/no response.

Table 3.1.1 **Common flowchart symbols**

When drawing flowcharts, arrows are used to join the symbols together. This shows the sequence of events that is taking place. These must always be drawn either horizontally or vertically. Symbols should also be drawn to a uniform size.

ACTIVITY

Draw a flowchart that shows the steps needed to complete an engineering project that you have been working on in the engineering workshop. Include quality control checks as appropriate.

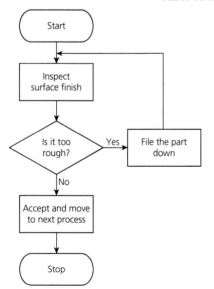

Figure 3.1.5 **Flowchart for a quality control process used to check the quality of surface finish**

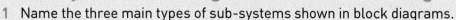

Check your knowledge and understanding

1 Name the three main types of sub-systems shown in block diagrams.
2 Describe the purpose of a circuit schematic.
3 Name the main types of symbols used when drawing flowcharts.

3.2

Mechanical systems

What will I learn?

By the end of this section you should have developed knowledge and understanding of:

→ how linkages work
→ how to calculate mechanical advantage, gear and velocity ratios
→ how mechanical systems can be used to change one form of motion into another
→ how cams and followers convert rotary motion to reciprocating motion
→ how gear trains and chain mechanisms can be used to transmit drive
→ how pulleys are used to transmit drive and increase mechanical advantage when lifting
→ the purpose of bearings and how they work.

The use and development of mechanical systems has driven engineering achievements since before the Industrial Revolution. The wind turbines that generate clean, sustainable energy and the engines that power our road vehicles are just two examples of how these systems influence our daily lives.

This section explains how a range of different types of mechanical systems function and how they are typically used. These include the use of linkages to change the direction of motion, gear trains to transmit drive and pulleys to ease the lifting of heavy objects.

You will learn about mechanical advantage and velocity ratios, how these are calculated and how to interpret the results of your calculations.

You will also learn about how different types of bearings are used to constrain motion and reduce friction in mechanical systems.

Linkages

Linkages are used to change the size of a force, the direction of motion and/or the type of motion. They are constructed by joining together links, rods or levers. These are connected via pivots that either allow or restrict their movement. A pivot is a point on which something balances or turns, and can be fixed or moveable. Different types of linkages affect movement and forces in different ways.

Types of linkage

Some of the most common types of linkage found in engineering applications are described in Table 3.2.1.

Linkage type	What it looks like	How it works
Reverse motion	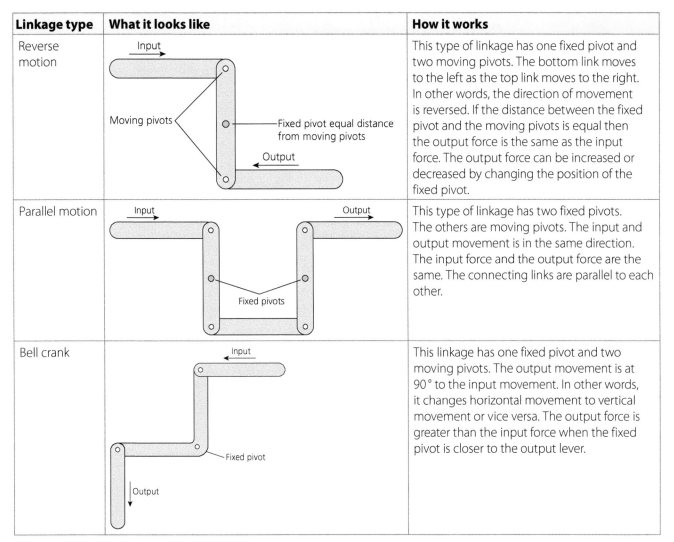	This type of linkage has one fixed pivot and two moving pivots. The bottom link moves to the left as the top link moves to the right. In other words, the direction of movement is reversed. If the distance between the fixed pivot and the moving pivots is equal then the output force is the same as the input force. The output force can be increased or decreased by changing the position of the fixed pivot.
Parallel motion		This type of linkage has two fixed pivots. The others are moving pivots. The input and output movement is in the same direction. The input force and the output force are the same. The connecting links are parallel to each other.
Bell crank		This linkage has one fixed pivot and two moving pivots. The output movement is at 90° to the input movement. In other words, it changes horizontal movement to vertical movement or vice versa. The output force is greater than the input force when the fixed pivot is closer to the output lever.

Table 3.2.1 **Types of linkage**

Mechanical advantage

Linkages can be used to provide **mechanical advantage**. This is the ability of a mechanism to move a large load with a small effort force. It is usually written as a number only without units.

Mechanical advantage (MA) $= \dfrac{\text{Load (Fb)}}{\text{Effort (Fa)}}$

In a linkage, the **effort** is the force at the input and the load is the force at the output. Force is measured in newtons (N).

MATHEMATICAL UNDERSTANDING

E15: Mechanical advantage calculation

Question:
What is the mechanical advantage of a linkage with an input force of 10 N and an output force of 30 N?

Solution:

$MA = \dfrac{Fb}{Fa}$

$MA = \dfrac{30}{10}$

$MA = 3$

Figure 3.2.1 **Oscillating motion example of a pendulum in a clock**

Conversion of motion

One of the main functions of mechanical systems is to change one type of motion into another.

Types of motion

The four main types of motion are:
- **linear motion**: motion in a straight line, for example, a train moving across a straight track or a person going down a slide
- **rotary motion**: motion that turns in a circle. An example of this is a wheel turning or a car driving around a roundabout
- **reciprocating motion**: motion that goes back and forth in a straight line, such as a person jumping up and down on a pogo stick
- **oscillating motion**: a swinging motion from side to side, for example, a pendulum in a clock or the movement of a swing in a park.

Converting motion types

Mechanical systems can be used to change the direction of motion in a system. A rack and pinion mechanism changes rotary motion to linear motion. The rack is a straight, bar shaped gear with teeth. The teeth on the circular pinion gear mesh with those on the rack. As the pinion gear turns, it results in one of two outcomes, depending on the design of the system. Either the rack will move in a straight line, or the pinion itself will experience linear motion. An example of the latter is a rack railway system.

Figure 3.2.2 **A rack and pinion mechanism**

A crank and slider mechanism can be used to convert rotary motion to reciprocating motion, and vice versa. To create reciprocating motion, the crank rotates and the connecting rod pushes the slider backwards and forwards. Piston engines in road vehicles are an example of the reverse. The piston acts as the slider which results in the wheel of the vehicle turning.

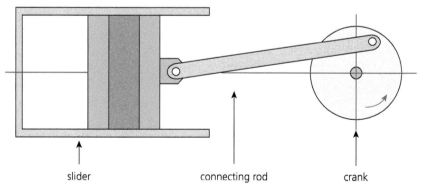

slider   connecting rod   crank

Figure 3.2.3 **A crank and slider mechanism**

Another way of converting rotary motion to reciprocating motion is through using a **cam and follower** mechanism. These are covered in more detail later in this section.

Gear trains

Gear trains transmit rotary motion and torque. They are made up of two or more spur gears that mesh together. A spur gear is a circular gear with straight teeth. The size and number of teeth on each of the gears affects the size of the output speed and the torque transmitted. Torque is a rotational force.

Simple gear trains

A simple two-spur gear train is shown below. A larger gear driving a smaller gear results in increased speed but decreased torque. A smaller gear driving a larger gear has the opposite effect. The driven gear will rotate in the opposite direction to the driver gear.

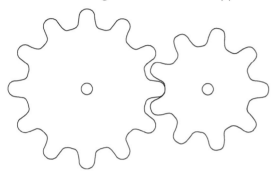

Figure 3.2.4 **A simple gear train**

The gear, or velocity, ratio of a simple gear train is calculated by dividing the number of teeth on the driven gear by the number of teeth on the driver gear.

$$\text{Gear ratio} = \frac{\text{Number of teeth on driven gear}}{\text{Number of teeth on driver gear}}$$

Chain and sprocket

An alternative to using gear trains to transmit rotary motion is the use of a **chain and sprocket** mechanism. In this mechanism, a series of links are joined together with steel pins to make the chain. The sprockets are toothed wheels. The chain fits over the sprockets, which are a set distance apart. As the driver sprocket turns so does the chain, also turning the driven sprocket.

This has the advantage over more complex gear trains as only two sprocket wheels and a chain are needed, thus saving cost. However, the chain can jump out of place or break if the system is not maintained correctly. Hence chain systems need to be tensioned, and additional

Figure 3.2.5 **A chain and sprocket mechanism**

spring-loaded mechanisms are often used. One of the most common applications of a chain and sprocket mechanism is as used on a bicycle pedal system.

As with a gear train, the velocity ratio can be worked out by dividing the number of teeth on the driven sprocket by the number of teeth on the driver sprocket. The length of chain or number of links is not important.

Cams and followers

Cams and followers turn rotary motion into reciprocating motion. The follower moves up and down as the cam rotates. The cam can be turned manually via a handle or automatically using a motor or other method.

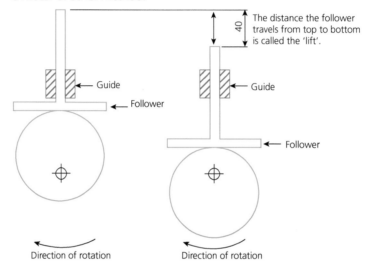

The distance the follower travels from top to bottom is called the 'lift'.

Guide

Follower

Direction of rotation

Figure 3.2.6 **An eccentric cam and follower mechanism**

The pattern of the movement produced is determined by the design of the cam. The follower will **dwell** (remain stationary) if the cam is round and the shaft is in the centre of the cam. Using different shaped cams and/or moving the shaft to a non-central position results in the follower rising (moving up) or falling (moving down).

Eccentric cams

Eccentric cams are circular but have an off-centre rotating shaft. With this type, the **rise** and **fall** produced is symmetrical.

Pear cams and egg-shaped cams

Pear cams or egg-shaped cams result in the follower remaining in dwell for half of the cycle. It is then pushed up as the point of the cam approaches. Finally, as the point passes, the follower falls and then dwells again.

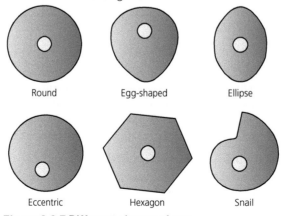

Round   Egg-shaped   Ellipse

Eccentric   Hexagon   Snail

Figure 3.2.7 **Different shapes of cam**

The use of cams and followers in engines

Cam and follower mechanisms are widely used in the internal combustion engines that power motor vehicles. They are typically used to control intake and exhaust valves.

A camshaft is a cylindrical rod that runs the length of the bank of cylinders in an engine. It has one cam fastened to it for each valve in the engine. As the camshaft rotates, the cams open and close the intake and exhaust valves by applying pressure to them at the appropriate times. This ensures that the valves operate in time with the pistons in the engine.

Figure 3.2.8 Camshaft and valves in a modern diesel engine

> ### STRETCH AND CHALLENGE
> Investigate how camshafts are used to control valves in internal combustion engines. Create a presentation of your findings to show to the class.

Pulleys

Pulley systems are used to reduce effort when lifting **loads** and to transfer power within a system. They transmit rotary motion. The pulleys are wheels that are placed a set distance apart. They can have grooves or teeth in them, or a flat top. In the case of grooved pulleys, a belt slots into the grooves. As the driver pulley turns so does the belt, thus transferring the rotary motion to the driven pulley. The belt is often made from rubber.

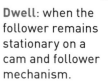

Figure 3.2.9 An electric motor driving a simple pulley system

The direction of rotation remains the same from the driver to the driven pulley, but the speed and torque can be adjusted by using different sizes of wheels. A larger pulley wheel driving a smaller wheel will result in increased speed but decreased torque. A smaller wheel driving a larger wheel results in the opposite effect. This therefore creates a similar result to gear trains.

The velocity ratio of a pulley system can be calculated by dividing the diameter of the driven pulley by the diameter of the driver pulley. The belt does not factor into the calculation.

$$\text{Velocity ratio} = \frac{\text{Diameter of the driven pulley}}{\text{Diameter of the driver pulley}}$$

>
>
> ### MATHEMATICAL UNDERSTANDING
>
> **Pulley system: velocity ratio**
>
> **Question**
> A driver pulley has a diameter of 30 mm and the driven pulley has a diameter of 150 mm. What is the velocity ratio of the system?
>
> **Solution**
> $$\text{Velocity ratio} = \frac{150}{30}$$
> $$= 5{:}1$$

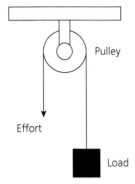

Effort

Pulley

Load

Figure 3.2.10 A pulley system used for lifting with a mechanical advantage of 1

An advantage of pulley systems is that they are cheaper and require less parts than gear train or chain systems. The belt however can break or slip, causing damage to the system.

Lifting loads with pulleys

Pulleys are commonly used to aid the lifting of heavy objects by reducing the effort. The simplest system of this type consists of a single pulley wheel and rope. The pulley wheel is attached high to a fixed spot, such as a roof support. The rope loops over the pulley. The load to be lifted is on one side and the effort is applied on the other. Although using the system is safer for the person applying the effort than not using the system, it only has a mechanical advantage of 1. This means that the effort needed to lift the load is not reduced by the pulley system.

One way of increasing the mechanical advantage is by making the pulley moveable and turning it upside down. In this next example, one end of the rope is attached to a fixed point above the pulley and a pulling force (effort) is applied to the other. This allows the pulling force to be halved giving a mechanical advantage of 2. For example, if the load was 60 N, the pulling force needed to lift it would be 30 N.

The design of this particular system does, however, mean that the effort would need to be applied from high off the ground, which might not be practical.

The example in Figure 3.2.11 can be modified so that the effort can be applied from the ground and still achieve a mechanical advantage of 2. To enable this to occur a fixed pulley must be added to the system as shown in Figure 3.2.12.

Additional pulleys can be added to the system to increase the mechanical advantage still further.

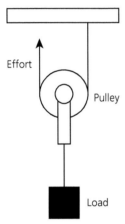

Effort

Pulley

Load

Figure 3.2.11 A pulley system with a mechanical advantage of 2

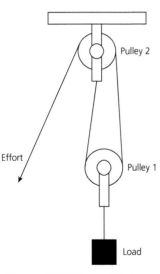

Pulley 2

Effort

Pulley 1

Load

Figure 3.2.12 A modified two-pulley system

MATHEMATICAL UNDERSTANDING

E15: mechanical advantage

Question
A person is applying a pulling force of 30 N to lift a load of 120 N. Calculate the mechanical advantage of the pulley system used.

Solution
$$MA = \frac{Load}{Pulling\ force}$$
$$MA = \frac{120}{30}$$
$$MA = 4$$

STRETCH AND CHALLENGE

Construct a pulley system capable of providing a mechanical advantage of 2. Use it to lift different loads and calculate the pulling forces required.

Bearings

Bearings are machine parts. Their role is to control motion and reduce friction between moving parts.

Plain bearings

Plain bearings are the simplest form of bearing. They provide a sliding contact with the other parts of the machine. There are several sub-types of plain bearing, including radial (sometimes called rotary), linear and thrust bearings. An example of a radial bearing in use is a shaft rotating in a hole.

Figure 3.2.13 A radial plain bearing

Rolling element bearings

Rolling element bearings consist of rolling elements placed between two bearings, known as races. They are designed to prevent sliding friction. The two most common sub-types are:

- ball bearings: these use spherical balls as the rotating elements
- roller bearings: these use cylindrical rollers.

Figure 3.2.14 A ball bearing

> ### KEY WORDS
>
> **Plain bearing**: a simple bearing that provides a sliding contact with the parts of a machine.
>
> **Rolling element bearing**: a bearing that consists of rolling elements placed between two races.

KEY POINTS

- Linkages are used to change the size of a force, the direction of motion and/or the type of motion.
- Mechanisms can be used to convert between linear, rotary, reciprocating and/or oscillating motion.
- Cams and followers convert rotary motion to reciprocating motion.
- Gear trains are used to increase or decrease speed and torque.
- The mechanical advantage of a pulley system is calculated by dividing the load force by the pulling force.
- Bearings constrain motion and reduce friction between moving parts.

Check your knowledge and understanding

1. State the function of a reverse motion linkage.
2. A mechanical system has an input force of 35 N and an output force of 140 N. Calculate the mechanical advantage of the system.
3. The driven gear in a gear train has 45 teeth and the driver gear has 135 teeth. Calculate the gear ratio of the system.
4. State the main purpose of a cam and follower mechanism.
5. Describe the two main functions of pulley systems.

Electrical systems

What will I learn?

By the end of this section you should have developed knowledge and understanding of:
- → the differences between alternating and direct current
- → the uses of mains electricity and different types of batteries as power supplies for engineered products and systems
- → the different types of input control devices that are commonly used in electrical systems, including switches and relays
- → the different types of output devices that are commonly used in electrical systems to create movement, sound and light.

Electricity is all around us and shapes our everyday lives. We depend on electrical power and electrical systems more than at any other time in history. The lighting, security and entertainment systems in our homes, for example, all rely on electricity being available at all times.

This section describes the relationship between the key concepts of voltage, current and resistance. It explores the different types of power supplies used in electrical systems, from mains power to low-voltage battery supplies. It investigates the difference between alternating current (AC) and direct current (DC) and the typical uses of each.

You will also learn about the different types of input control and output devices used in electrical systems. These include switches, relays and outputs that create light, sound and movement.

KEY WORDS

Alternating current: current that changes direction periodically.

Mains electricity: electricity supplied via plug sockets at 230 V.

Electric current

An electric current is a flow of electric charge through a conductive medium, such as a wire. The SI unit of current is the ampere, or amp. Ohm's Law states that the current flowing through a resistor is directly proportional to the voltage across the resistor. This is represented using the formula:

$$\text{Current (I)} = \frac{\text{Voltage (V)}}{\text{Resistance (R)}}$$

Ohm's Law and the relationship between voltage, current and resistance is described in more detail in Section 4.1.

There are two types of electrical current that are used:
- alternating current (AC)
- direct current (DC).

Alternating current

Alternating current (AC) changes direction periodically: 50 times a second in the case of **mains electricity** in the home. The voltage level reverses along with the current. AC usually has an oscillating voltage that appears graphically as a sine wave.

Mains electricity is an AC supply in the United Kingdom. This is because it is relatively straightforward to transport it across long distances. The AC frequency is usually set at 50 Hz. This means that it changes direction and back again 50 times per second.

One common application of AC is the powering of electric motors, such as those found in washing machines, tumble dryers and dishwashers.

Direct current

Direct current (DC) flows in one direction only. This can also be shown graphically. In this example, a 1.5 V DC supply is being provided.

Many digital electronic products run using DC, which can be provided by batteries and cells. These products include mobile phones, tablet computers and e-readers. The main advantages of DC electricity are its reliability and efficiency.

Some digital devices use a converter to turn AC into the DC supply that is required. This process is known as rectification. This then allows the devices to be plugged into a mains socket. An example of this is a modern flat-screen television. USB adaptors can be used to convert AC from a mains socket to DC. An example application of this is the charging of mobile phone batteries.

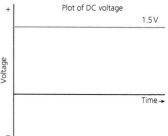

Figure 3.3.1 **Alternating current sine wave**

Figure 3.3.2 **Graph of direct current**

> **STRETCH AND CHALLENGE**
> Explain, in detail, the differences between AC and DC. Use graphs to support your response.

Power supplies

Electrical devices must be powered for them to work. The two most common sources of power for most electrical and electronic systems are mains electricity and batteries.

Mains electricity

Mains electricity is supplied at 230 V AC in the United Kingdom. Electrical outlets, or sockets, allow devices to make use of the electricity provided using a three-prong (live, neutral, earth) plug. The plug is usually connected to the appliance via flexible cable. A 'step down' transformer can be used to reduce the voltage to a lower level if the appliance requires this.

The advantages of using mains electricity are that it is relatively cheap and available in the vast majority of homes and business properties in the UK. However, appliances that use mains electricity must be physically plugged into a socket. This can present a problem if an appliance must be portable or needs to be used where there is no access to mains supply. In addition, there may only be a limited number of sockets available in a particular room or space.

Due to the high voltages involved, mains electricity can be extremely dangerous if used incorrectly or unsafely. Contact with live parts can result in electric shocks and burns. Faulty mains equipment or wiring can also cause fires, or even explode during use. When using mains powered devices it is therefore important to follow safe practices.

Batteries

Batteries consist of two terminals (an anode and a cathode) and an electrolyte. They convert chemical energy into electrical energy. When a **battery** is connected as the power supply to a circuit or device, chemical reactions occur on the electrodes. This results in a flow of electrical energy to the circuit. The circuit symbol for battery is shown in Figure 3.3.4

KEY WORDS

Direct current: current that flows in one direction only.

Battery: a device that converts chemical energy into electrical energy.

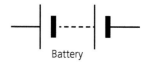

Figure 3.3.3 **A UK electrical socket**

Battery

Figure 3.3.4 **Battery circuit symbol**

Figure 3.3.5 Different battery formats (from left to right: D, C, AA, AAA and PP3)

Batteries come in a range of different sizes, shapes and voltages including AAA, AA, C, D (1.5 V) and PP3 (9 V) formats.

Examples of different types of batteries include alkaline, zinc carbon and rechargeable batteries. Single-use batteries are thrown away once they are no longer useful, whereas rechargeable batteries can be recharged hundreds or even thousands of times. Rechargeable batteries are more expensive to buy than single-use batteries, but often work out cheaper over time due to the number of times that they can be re-used. Using rechargeable batteries also reduces the amount of waste that has to be thrown away.

Batteries are useful for powering electronic products or devices that are to be used where there is no access to mains electricity, for example, a portable DVD player to be used by children on long car journeys. They can also be placed in a 'pack' to create a specific voltage. For example, three 1.5 V AA batteries can be connected in series to create a 4.5 V power supply. Pre-manufactured cases can be purchased for this purpose.

Input control devices

Switches

Switches are used to either 'make' (allow current to flow through) or 'break' (do not allow current to flow through) a circuit. A common use of switches is to turn the power supply to a circuit on or off.

Figure 3.3.6 Circuit symbol for a switch

Switches are usually described by their methods of operation. These include push-to-make, push-to-break, slide, micro-, rocker, tilt and reed switches.

Figure 3.3.7 Rocker switch

The type of **switch** used will depend on the needs of the system. Micro-switches are very sensitive and are hence suitable for detection control applications. Reed switches consist of two contacts, or reeds, inside a glass tube that touch when exposed to a magnetic field. They are commonly used in security applications, such as window alarms.

Relays

Relays are electrically operated switches. When current flows through the internal coil, a magnetic field is created. This attracts a lever which changes the switch contacts by pushing them together. This completes the circuit. The contacts return to their open state when the current is switched off.

Figure 3.3.8 Reed switch

Relays are used to switch a large load current from a much smaller control current. They allow one circuit to switch a second, completely separate circuit. For example, switching a 230 V mains circuit from a low-voltage battery-powered circuit. Some types of relay can also be used in latching applications.

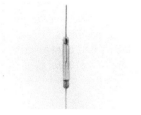

Figure 3.3.9 Micro-switch

Figure 3.3.10 An electromagnetic relay

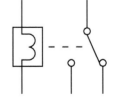

Figure 3.3.11 Circuit symbol for a relay

Output devices

Output devices can be used to provide light, sound or movement. As with inputs, the outputs selected depend on the requirements of the system.

Motors and solenoids

Motors and **solenoids** are the two main ways of providing movement in an electrical system. Motors convert an electrical signal, such as current, into rotary motion. They can be powered by either AC or DC sources. They can also be used in reverse as power generators.

Figure 3.3.12 **Circuit symbol for a motor**

An example of a low-cost motor appropriate for most engineering prototyping work in schools is a DC solar motor.

A servo motor can be used when precision speed or positioning is required, such as in CNC equipment. Stepper motors can be used when a full rotation needs to be divided into equal steps, such as in camera positioning and robot arms.

Solenoids convert electrical energy into mechanical movement through the use of electromagnetism. This is usually in the form of linear motion but can also be rotary. They are often used in automated locking and clamping systems.

STRETCH AND CHALLENGE

Prepare a detailed technical report explaining how motors and solenoids work. Your report should focus on the different types and the output signals that they produce. You should also include common applications of each, circuit schematics and CAD simulations to support your responses.

ACTIVITY

Draw and model a simple circuit showing how a switch can be used to turn a lamp on or off.

Figure 3.3.13 **A DC solar motor**

Buzzers and bells

Buzzers and **bells** are used to create sound. Buzzers use an internal oscillator to produce sounds at different frequencies when current flows through them. The frequency depends on the design and voltage rating of the buzzer, which is typically 6 V or 12 V for the types used in schools. They are commonly used in doorbells, quiz buzzers and alarm systems.

Bells can also be used to create a sound output through the use of an electromagnet. When current flows through a bell it produces a continuous clanging sound. A good example of this is a school bell. Bells are less commonly used nowadays, as many have been replaced with fully electronic sound output devices.

Figure 3.3.14 **Circuit symbol for a buzzer**

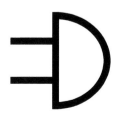

Figure 3.3.15 **Circuit symbol for a bell**

Lamps

Lamps are output devices used to create light. Filament lamps consist of a thin coil of wire, known as the filament, within a glass outer casing. When current passes through the filament it heats up, thus producing light.

Lamps are used in a wide range of lighting applications including torches, task lights and indicator lights. They are however being gradually replaced by more energy efficient devices, such as light emitting diodes (LEDs).

When selecting a lamp for use in an engineering project it is important to pay attention to the voltage rating, current rating and fitting type required. The voltage rating is the supply voltage that is needed for normal brightness to be achieved. Using too high a supply voltage can damage or destroy the lamp. Typical lamp fittings include screw, centre contact and bayonet caps.

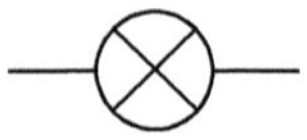

Figure 3.3.16 **Lamp circuit symbol**

KEY WORDS

Switch: an input device that is used to either 'make' or 'break' a circuit.

Relay: an electrically operated switch used to switch a large load current from a much smaller control current.

Motor: a device that converts an electrical signal into rotary motion.

Solenoid: an output device that converts electrical energy into mechanical movement through the use of electromagnetism.

Buzzer: an output device that uses an internal oscillator to produce sound.

Bell: an output device that creates sound using an electromagnet.

Lamp: an output device that produces light as a result of heat when current flows through it.

KEY POINTS

- Ohm's Law states that current is directly proportional to voltage, and inversely proportional to resistance.
- Direct current flows in one direction only, whereas alternating current changes direction periodically.
- Switches are used to 'make' or 'break' a circuit.
- Motors and solenoids are output devices used to convert electrical signals into movement.
- Buzzers and bells are output devices used to produce sound.

Check your knowledge and understanding

1 State the formula for Ohm's Law.
2 Explain the difference between alternating current and direct current.
3 Describe the main function of a relay.
4 State the type of output signal produced by lamps.

3.4 Electronic systems

What will I learn?

By the end of this section you should have developed knowledge and understanding of:

→ analogue and digital signals and the differences between them
→ how input sensors can be used to detect changes in light and temperature levels
→ how process blocks produce functions such as timing, counting, comparison of input signals and logic
→ how peripheral interface controller (PIC) microcontrollers can be used to replace discrete components and integrated circuits (ICs)
→ how transistors and field-effect transistors (FETs) are used in driver circuits
→ the functions of resistors, capacitors and diodes in circuits
→ how monitoring and control processes can be programmed.

Electronic systems add functionality and intelligence to engineered products. From mobile smartphones to complex security, lighting and heating systems, it is hard to imagine the modern world without them.

This section describes the types of signals that electronic system blocks receive and output. It looks at how input sensors are used to detect environmental changes and how process sub-systems create functions such as timing, logic and counting. It will also explain the use and function of resistors, capacitors and **diodes** in electronic circuits.

You will learn how programmable devices can be used to replace discrete components within a system and how the signals from these can be boosted with driver circuits. You will also learn how these signals are then outputted using components such as LEDs and piezo sounders.

Analogue and digital signals

Electronic systems and sub-systems collect, transmit, alter and output both **analogue** and **digital signals**.

Analogue signals are signals that vary and change continuously over time. They can take any value within a given range. Examples of analogue signals include sound waves, light levels and continuously varying voltages or electrical currents.

Value

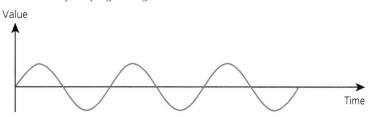

Time

Figure 3.4.1 **Example of an analogue signal waveform**

Digital signals are sent as pulses of information that are either high (1) or low (0). They can carry more information per second than analogue signals and maintain their quality better. An example of the use of digital signals is the information contained on CDs and MP3 files.

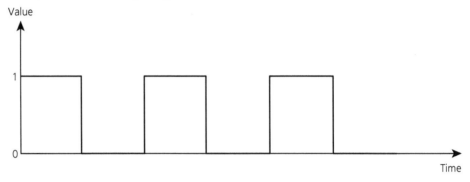

Figure 3.4.2 **Example of a digital signal waveform**

Figure 3.4.3 **A light-dependent resistor (LDR)**

Figure 3.4.4 **Circuit symbol for an LDR**

Figure 3.4.5 **A thermistor**

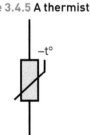

Figure 3.4.6 **Circuit symbol for a thermistor**

Sensor inputs

Sensors are used as inputs in electronic systems. They allow systems to gather information about the environment around them, for example, detecting changes in light or temperature levels. This information is then changed into an electrical or electronic signal that can be understood and acted on by the process sub-system.

Sensor inputs can be analogue or digital. Analogue sensors output a continuous number of different values, whereas digital sensors output either a high or a low signal. Two electronic components that are commonly used in sensing applications are light-dependent resistors (LDRs) and thermistors. These are examples of analogue sensors.

Light-dependent resistors

A light-dependent resistor (LDR) changes the light level that is detected into a resistance. It is therefore a special type of variable resistor. Its resistance decreases as the brightness increases.

LDRs are relatively cheap to buy and readily available for use in engineering projects. An example use of an LDR is in a sensing circuit for a garden lamp that lights up when it gets dark.

Thermistors

Like LDRs, thermistors are special types of variable resistors. They change the temperature level that is detected into a resistance. In the vast majority of thermistors, the resistance decreases as the temperature increases.

An example use of a thermistor is in a sensing circuit for a system that needs to keep a room at a specific temperature, such as a greenhouse.

> **ACTIVITY**
>
> Make a list of possible engineering applications that could make use of light and/or temperature sensors. Explain how the sensors would work as part of the larger system to achieve the desired outcome.

Process devices

Process devices are often thought of as the 'brain' of an electronic system. They work by responding to the electrical signal sent from the input stage and changing it in some way. This altered signal is then used to control the output stage(s) of the system.

Process devices usually take the form of an integrated circuit (IC), which is a complete small circuit contained on a microchip. The rest of the system is then built around this chip, with various components working alongside it to achieve the desired functions.

The most common types of processing devices are used to achieve functions such as timing, counting, comparison of different signals and logic.

The next few pages look at non-programmable methods of achieving these functions. Programmable devices are covered later in this section.

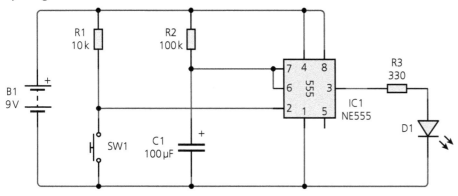

Timers

Timer circuits are used to turn an output high (on) or low (off) for a certain amount of time. This is usually in response to an input signal of some kind, such as a push switch being pressed or a sensor detecting a change in the environment. One way of achieving a timing function is by using a 555 monostable circuit.

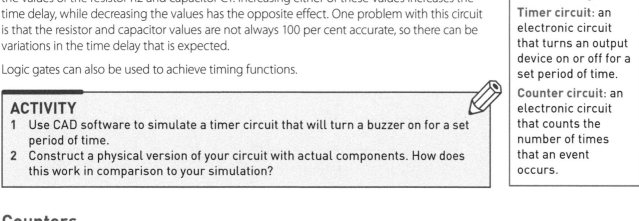

Figure 3.4.8 **A 555 monostable timing circuit**

The example shown in Figure 3.4.8 would light up the LED labelled as D1 for approximately 11 seconds when the switch is pressed. The length of the time delay can be altered by changing the values of the resistor R2 and capacitor C1. Increasing either of these values increases the time delay, while decreasing the values has the opposite effect. One problem with this circuit is that the resistor and capacitor values are not always 100 per cent accurate, so there can be variations in the time delay that is expected.

Logic gates can also be used to achieve timing functions.

> **ACTIVITY**
> 1   Use CAD software to simulate a timer circuit that will turn a buzzer on for a set period of time.
> 2   Construct a physical version of your circuit with actual components. How does this work in comparison to your simulation?

Counters

Counter circuits are used to count the number of times an event happens, for example, each time a point is scored in a sports event. Often each count is triggered by a switch being pressed, although sensing circuits can also be used for this purpose. An example of a commonly used counting circuit is a decade counter (see Figure 3.4.9).

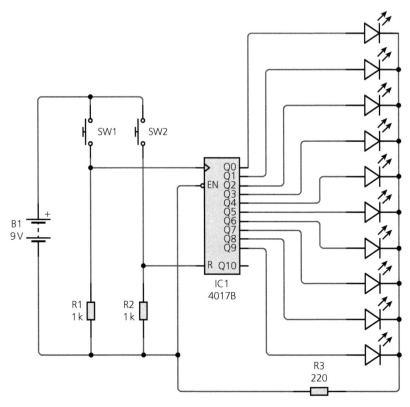

Figure 3.4.9 **A decade counter circuit**

The circuit in Figure 3.4.9 is based around the 4017B IC. When the switch SW1 is pressed a low to high pulse is sent to the clock input on the IC and the first LED is lit. Every time the switch is subsequently pressed the previous LED turns off and the next LED lights. The switch SW2 is used to reset the counting sequence back to the start.

Comparators

Comparator circuits compare two input voltage signals and decide which one is the greater. An operational amplifier (op-amp) is often used as a process block to achieve this. The results can then be indicated using one or more output devices, such as LEDs, lamps and buzzers.

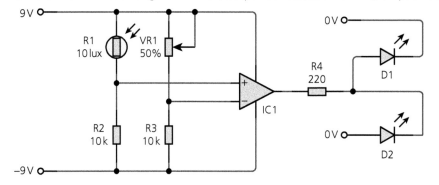

Figure 3.4.10 **A comparator circuit**

The op-amp (IC1) has two inputs:
● an inverting input (−)
● a non-inverting input (+).

It compares the inverting input voltage with the non-inverting input voltage. If the non-inverting input voltage is higher, the op-amp output goes high (LED 1 lights up). If the inverting input voltage is higher the output goes low (LED 2 lights up). In this instance, an LDR (R1) and a variable resistor (VR1) are being used to adjust the input voltages.

The sensors used can be changed to detect different environmental changes for different applications. For example, using a thermistor could allow the comparator to indicate whether a space is above or below a certain temperature. This would be useful in situations where temperature must be precisely controlled, such as a food storage area.

Logic gates

Logic gates respond to and output digital signals; that is, signals that are either high (1) or low (0). The main three types of logic gate that you need to be aware of are:
- NOT gates
- AND gates
- OR gates.

Truth tables are used to show what the output signal is for different combinations of possible input signals.

NOT Gate

With a NOT gate, the output signal Q is the opposite of the input signal A; that is, the output signal is *not* the same as the input signal. It is sometimes described as an inverter.

A	Q
0	1
1	0

Figure 3.4.12 **NOT gate truth table**

NOT gates can be used to turn an output device off when a switch is turned on, for example, an emergency stop button on a machine.

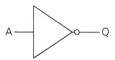

Figure 3.4.11 **Circuit symbol for a NOT gate**

AND Gate

AND gates only produce an output signal of 1 when both input signals are 1; that is, the output signal Q is only 1 when both input A *and* input B are 1.

A	B	Q
0	0	0
0	1	0
1	0	0
1	1	1

Figure 3.4.14 **AND gate truth table**

An example of the use of an AND function is in a lift or elevator system. The lift should only go up to the next floor when both the floor button has been pressed and a sensor has detected that the door has closed.

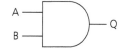

Figure 3.4.13 **Circuit symbol for an AND gate**

OR Gate

OR gates produce an output signal of 1 when either input signal is 1; that is, the output signal Q is 1 when either input A *or* input B is 1.

A	B	Q
0	0	0
0	1	1
1	0	1
1	1	1

Figure 3.4.16 **OR gate truth table**

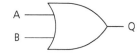

Figure 3.4.15 **Circuit symbol for an OR gate**

Process devices

An example of the use of an OR function is in a simple home alarm system. The alarm should sound when either the door or the window sensor detects a break in.

Programmable devices

Programmable devices can be used to perform more complex operations than discrete integrated circuits. They can also replace them for simple functions such as timing, counting and comparing signals. A programmable device must have a program downloaded onto it to perform the required functions.

Figure 3.4.17
Microcontrollers in 8-, 16- and 28-pin formats

Microcontrollers

The programmable device most widely used in engineering projects is a **microcontroller**. This is basically a small computer on an integrated circuit. It has different pins, or ports, for the connection of input and output devices.

Peripheral interface controller

A **peripheral interface controller (PIC)** is a type of microcontroller that is common in both schools and industry. The simplest PIC has 8 pins, but more complex versions can have up to 40. For most GCSE level work, an 8- or 14-pin microcontroller should suffice.

PICs offer a number of advantages over discrete integrated circuits:
- they can be programmed and reprogrammed thousands of times to perform a range of different functions, making them very flexible and adaptable
- one microcontroller can replace a number of discreet components or ICs, saving space on a circuit
- multiple input and output devices can be connected to them, allowing them to be used in very complex systems.

PICs can, however, be expensive, so for simpler functions it is sometimes more cost effective to use discrete components. Access to programming software and hardware is also required, which can also increase costs.

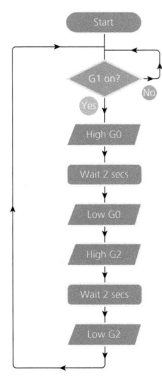

Figure 3.4.18 A PIC microcontroller program for a timer with one input and two outputs

Programming PICs

PICs can be programmed using block-based editors, flowchart software or text-based programming languages (such as BASIC or C). When using flowchart software, the correct symbols must be used. A list of flowchart symbols used for microcontroller programming is shown in the Engineering symbols section of this book (p. 218).

An example of a flowchart-based PIC microcontroller program, written using commercially available software, is shown in Figure 3.4.18.

In the example in Figure 3.4.18, when input port G1 detects a high signal, output G0 goes high (on) for two seconds. It then goes low (off). Output G2 then goes high for two seconds before going low.

A switch could be connected to the microcontroller as the input device, and lamps or LEDs connected as the outputs to create a flashing light sequence when the switch is turned on.

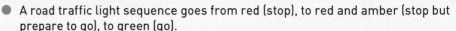

STRETCH AND CHALLENGE
- A road traffic light sequence goes from red (stop), to red and amber (stop but prepare to go), to green (go).
- Write a microcontroller program that would correctly operate a set of traffic lights as described above. Your program should have appropriate time delays built in between each light change in the sequence.

3.4 Electronic systems

Analogue to digital conversion (ADC) in a programmable device

Microcontrollers only understand digital signals, such as those from a switch being turned on, or a high/low signal from a digital sensor. However, many input devices produce analogue signals. As a result, most modern microcontrollers have an inbuilt analogue to digital converter (ADC) that can convert an analogue voltage to a digital value. To make use of this, the analogue input device must be connected to an input port with ADC capability.

Programming for an analogue input

Most flowchart programming software has a dedicated analogue input command that can be used to read a signal from an analogue sensor, and convert it to a digital output signal with the aid of a simple supporting program.

The program shown in Figure 3.4.19, written using commercially available software, reads the values coming from a sensor connected to analogue input port A1 of a PIC microcontroller. If the analogue signal falls outside of the given range (this can be adjusted), output port G0 goes high (on). Otherwise the port stays low (off).

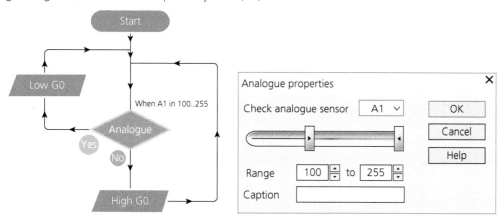

Figure 3.4.19 **Analogue input flowchart program**

Interfacing components

Interface components, or **drivers**, boost the output signal going from the process block of an electronic system to the output stage. This is necessary because some output devices require a larger current or voltage than can be provided by the process block alone. This is particularly the case with PIC microcontrollers as they only output small current values.

Transistors

Transistors are current amplifiers. They can provide enough current for outputs that require a small to medium current boost, such as lamps and buzzers. A transistor has three legs called the base, collector and emitter. The most common type of transistor used in driver circuits is an NPN bipolar transistor.

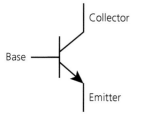

Figure 3.4.21 **Circuit symbol for an NPN transistor**

Figure 3.4.20 **A transistor**

An example of a transistor driver circuit is shown in Figure 3.4.22.

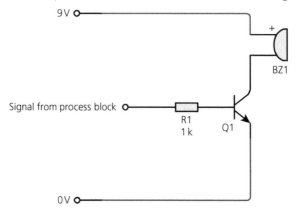

Figure 3.4.22 Transistor driver circuit

In the example in Figure 3.4.22, when the voltage between the base and the emitter is approximately 0.6 or greater, a small current flowing into the base will cause a larger current to flow from the collector to the emitter. This then allows the buzzer (BZ1) to sound.

Field effect transistors (FETs)

Field effect transistors (FETs) are voltage amplifiers. Whereas transistors are analogue devices, FETs operate digitally. A FET also has three legs but these are called the gate, drain and source. FETs can provide an output signal high enough for motors and solenoids to operate correctly. They have a very high input resistance which makes them useful as drivers for low-powered digital process devices, such as logic gates.

An example FET driver circuit for a motor is shown in Figure 3.4.24.

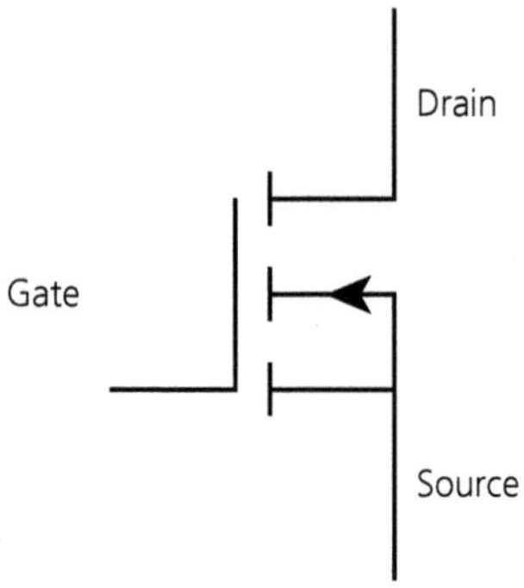

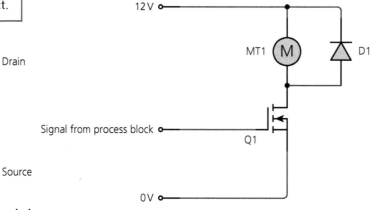

Figure 3.4.24 FET driver circuit for a motor

When the voltage at the gate is 2V or more the FET switches on and the motor turns.

Drain

Gate

Source

Figure 3.4.23 Circuit symbol for a FET

Output components

Output components turn an electronic signal, such as a voltage or current, back into a real-world signal, such as light, sound and movement.

Light-emitting diodes (LEDs)

Light-emitting diodes (LEDs) are output devices that produce light when current flows from the anode to the cathode. The long leg (round side) of the LED is the anode, whereas the short leg (flat side) is the cathode.

LEDs are often used as indicator lights, such as to show whether a device is powered on or off. Higher power 'hi-bright' versions are used increasingly in lighting applications as they use far less energy than incandescent lamps. They come in a range of colours including white. Flashing versions are also available.

LEDs can be damaged by too much electrical current, so they usually require a protective resistor connecting in series with them. However, they also have internal resistance approximating to 2 V forward drop.

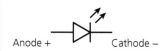

Figure 3.4.25 **A light-emitting diode (LED)**

MATHEMATICAL UNDERSTANDING

E12: Ohm's law
The required value of a protective resistor for an LED can be calculated using Ohm's Law.

Question
An LED requires 20 mA (0.02 A) to light and has a voltage drop of 2 V across it. If the supply voltage is 9 V, what is the required protective resistor value?

Solution
$V = I \times R$

So, $R = \dfrac{V}{I}$

$V = 9 - 2 = 7$

$R = \dfrac{7}{0.02}$

$R = 350 \, \Omega$

As this is an unusual value for a resistor, a 330 Ω resistor could be used instead.

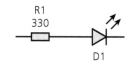

Figure 3.4.26 **Circuit symbol for an LED**

Figure 3.4.27 **LED with protective resistor connected in series**

7-segment display

A **7-segment display** is a package of LEDs arranged so that numbers ranging from 0 to 9 can be displayed. Each of the seven LEDs contained in the package is used as a 'segment' of the number to be shown. Different combinations of the segments light up to show the overall number. Some 7-segment displays have an eighth LED which is used to display decimal points. This is useful when two or more are used together to show larger or more complex numbers.

There are two types of 7-segment display. The segments on common cathode displays light when they receive high (logic 1) input signals. The segments on common anode displays light when they receive low (logic 0) signals. When including a 7-segment display in a circuit, a suitable driver is needed. A typical example is a 4511 binary-coded decimal (BCD) IC. 7-segment displays can also replace the standalone LEDs used in decade counter circuits.

Figure 3.4.28 **A 7-segment display**

Piezo sounder

Piezo sounders convert an electrical signal into sound using the piezo-electric effect. They are more versatile than buzzers (see Section 3.3) as they can produce a range of different tones. This makes them ideal for playing musical tunes, such as in a novelty greeting card or child's toy. A microcontroller can be used to provide the required signals for the different tones and most PIC programming software has commands specifically designed for this purpose.

Piezo sounders are supplied in cased or uncased formats.

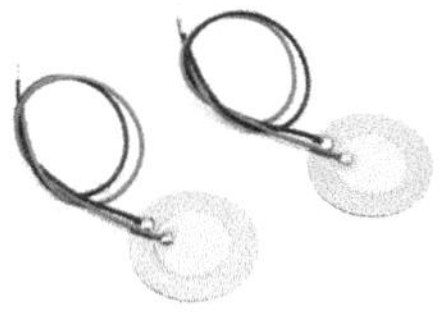

Figure 3.4.29 **A pair of uncased piezo sounders**

Discrete components within a circuit

Not all components in a circuit are classified as input or output devices, but they still have an important role to play in ensuring that a system works as expected. Resistors, diodes and capacitors are known as passive components as they do not require a source of energy to perform their stated functions.

Resistors

The purpose of a **resistor** is to reduce the flow of current in a circuit, for example, protecting an LED by reducing the amount of current flowing through it. The resistance value of a resistor is given in ohms, usually represented by the Greek symbol Ω. The higher the value in ohms, the greater the resistance of the resistor. Resistor values in the thousands are presented using a k (kilo), and in the millions using an M (mega). For example, $1000\,\Omega$ is simplified to $1\,k\Omega$. $1\,000\,000\,\Omega$ is simplified to $1\,M\Omega$.

When showing resistor values on a circuit diagram, a slightly different method is used. The Ω symbol and any decimal points are not shown. These are instead replaced by R (multiply by 1), K (multiply by 1000) or M (multiply by 1 000 000) as appropriate. This is best explained using examples:

- $330\,\Omega$ would be written as 330R, or even sometimes just 330
- $1000\,\Omega$ would be written as 1K
- $2700\,\Omega$ would be written as 2K7
- $1\,000\,000\,\Omega$ would be written as 1M

There are two main types of resistors used in circuits:
- fixed resistors
- variable resistors.

Fixed resistors

Fixed resistors have a set value within a tolerance. In older resistor designs, this is often between 5 per cent and 10 per cent of the stated value. More modern resistors can be produced to much tighter tolerances – typically around 2 per cent.

Only certain values of fixed resistor are readily available so it is sometimes necessary to combine resistors in series to achieve the required resistance for an application.

MATHEMATICAL UNDERSTANDING

To calculate resistor values in series you simply add up the value of each individual resistor.

Total resistance in series = R1 + R2 + R3 ...

Question
If R1 is $1\,k\Omega$, R2 is $10\,k\Omega$ and R3 is $100\,\Omega$, what would be the total resistance?

Solution
$1000 + 10\,000 + 100 = 11\,100\,\Omega = 11.1\,k\Omega$

Variable resistors

Variable resistors can be adjusted using either a small screwdriver hole or a spindle with an operating knob, depending on the type used. Pre-set potentiometers are usually set to a value and then left, whereas those with an operating knob are designed to be adjusted regularly. Examples of this type include a volume control on an electric guitar or radio.

Light-dependent resistors (LDRs) and thermistors are special types of variable resistor whose resistance changes with light and temperature levels respectively. This makes them ideal for use as input sensors.

Diodes

Diodes are components in a circuit that allow current to flow in one direction only. A diode has two legs called the anode and the cathode. Current flows from the former to the latter only. The cathode is usually marked with a silver or black line, or a negative (–) symbol.

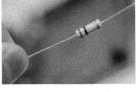

Figure 3.4.30 **A fixed carbon film resistor**

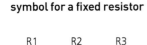

Figure 3.4.31 **Circuit symbol for a fixed resistor**

R1   R2   R3

Figure 3.4.32 **Three fixed resistors connected in series**

Figure 3.4.33 **Different types of variable resistors**

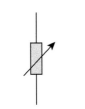

Figure 3.4.34 **Circuit symbol for a variable resistor**

Small-signal diodes are usually used in low-current applications, such as TVs, radios and mobile phones. Larger, power diodes are used in rectifying circuits. These are circuits that convert alternating current (AC) to direct current (DC). A common example of this is the bridge rectifier circuit.

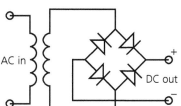

Figure 3.4.35 **Power diodes**

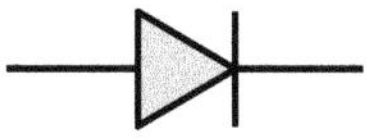

Figure 3.4.36 Circuit symbol for a diode

Figure 3.4.37 A bridge rectifier circuit

Capacitors

Capacitors store electrical charge. Their capacitance is measured in farads (F). A standard 1 F capacitor used to be very large in size, so for practicality most capacitor values used to be in micro farads, or µF. Super capacitors were developed to get around this problem. Their internal design allows them to have much higher capacitance values but in a small package, making them suitable for use as power supplies.

Capacitors can be either polarised or non-polarised. Polarised capacitors must be connected the correct way around in a circuit. They have a positive and negative leg. The negative leg is shown with a minus (–) sign on the capacitor casing. This leg is also shorter than the positive leg.

Non-polarised capacitors can be connected any way around in a circuit. These are much smaller physically and have less available capacitance values.

Figure 3.4.38 **Circuit symbol for a polarised capacitor**

Figure 3.4.39 **An electrolytic (polarised) capacitor**

Figure 3.4.40 Circuit symbol for a non-polarised capacitor

Figure 3.4.41 A ceramic disc (non-polarised) capacitor

Simple programming for monitoring and control processes

Monitoring and control processes are designed to make sure that engineered products are produced to a level of consistency, economy and safety that would not be possible using manual processes alone. These processes must be programmed so that they take place automatically and with a minimum of human interaction.

An example of this is a pick and place machine used in the production of electronic circuits. As many modern commercial circuit boards have thousands of components on them, it simply would not be possible to manually place them onto the circuit board with the speed, accuracy and precision required. A pick and place machine automatically selects the components needed and places them into the correct position on the circuit board using a robot arm. The information needed to do this is programmed into the machine. The flowchart in Figure 3.4.42 describes how a pick and place system operates.

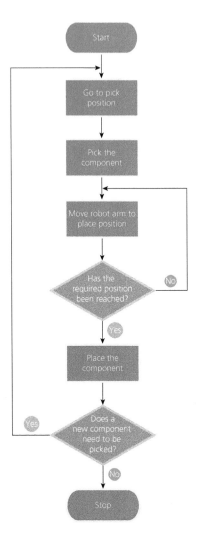

Figure 3.4.42 **Flowchart for a pick and place system**

ACTIVITY

Identify an industrial engineering monitoring and control process that you have studied. Create a flowchart showing how this process is carried out.

KEY WORDS

Discrete component: an individual or separate component.

Resistor: a component that reduces the flow of current in a circuit.

Capacitor: a component that stores electrical charge.

KEY POINTS

- Analogue signals vary continuously within a given range, whereas digital signals are either high (1) or low (0).
- Input sensors take real-world signals, such as light or temperature levels, and change them into electrical signals.
- Process blocks act as the 'brain' of a system. Examples include counters, timers, logic gates and comparators.
- Microcontrollers are programmable devices that can replace discrete components.
- Drivers boost the electrical signal from the process block of a system to the output block.
- Light-emitting diodes (LEDs) usually require a protective resistor connected in series.
- 7-segment displays produce a numerical output from 0 to 9.
- Resistors, diodes and capacitors are examples of passive discrete components.

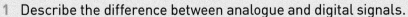

Check your knowledge and understanding

1 Describe the difference between analogue and digital signals.
2 Name an input sensor that can be used to detect changes in temperature levels.
3 Explain how a comparator circuit works.
4 State what PIC stands for.
5 State the function of a capacitor in a circuit.

3.5

Structural systems

What will I learn?

By the end of this section you should have developed knowledge and understanding of:

→ static and dynamic load forces

→ the stresses that can lead to a structural system buckling, bending or being pulled apart

→ the differences between space frame and monocoque structures.

We are surrounded by structures. From the homes we live in, to the aircraft we fly in, structural systems ensure that engineered products can cope with the forces and stresses that will be placed on them every day of their use.

This section explains the types of forces and stresses that can cause structures to deform and fail. It looks at how these are resisted, the different types of structural systems that can be used and their advantages and disadvantages.

You will learn about static and dynamic loads, stresses that cause structures to bend and buckle and the difference between space frame and monocoque structures.

Structural systems

The purpose of a structural system is to resist loads and forces that could otherwise cause the main structure to deform or fail. A classic example is the steel frame 'grid' design that made the development of the skyscraper possible.

Figure 3.5.1 **A structural system constructed using a steel frame**

There are various different types of structural systems. This section will look at:

● space frame structures

● monocoque structures.

Load and stress

In order to understand how structural systems work, it is important to learn about the forces that act on them. Loads are forces that place stresses on a structure and can cause it to deform if it is not capable of resisting them sufficiently.

Static and dynamic loads

Static loads, also known as dead loads, are forces that are constant. That is, they stay the same for a long period of time. For example, the weight of a building is a simple example of a static load.

Figure 3.5.2 **A bridge is subjected to both static and dynamic loads**

Dynamic loads, or live loads, are not constant and therefore vary over time. For example, in addition to the static load of its weight, a bridge could be subjected to several dynamic loads, such as the weight of cars moving across it or the wind blowing against it.

Calculations of loads, stress and strain are covered in Section 4.1.

Structural stresses

Structural stresses can be tensile or compressive. **Tensile stresses** occur as a result of pulling or stretching. They can cause structural components to become elongated or even pull them apart.

Compressive stresses occur as a result of pushing forces. They attempt to reduce the size of structural components. Compressive stresses can cause buckling within structural systems.

Bending can occur when there is a combination of tensile and compressive stresses present. Torsion occurs as a result of twisting forces.

Structural failure takes place when a structure is stressed beyond its limits. Structures must therefore be designed so that they can sufficiently resist the loads and stresses that are to be placed on them. Engineers use structural analysis techniques, including modelling and calculations, to ensure that this will be the case. Failure to do this can be catastrophic. A large bridge collapsing could cause the deaths of many people and cost millions of pounds to rebuild or repair.

ACTIVITY

Discuss the forces and stresses acting on a bridge or other structure. Are the loads static, dynamic or both? What stresses is the structure subject to? How does the design of the structure resist the loads and stresses acting on it?

Space frame structures

A **space frame structure** is constructed from struts that lock together in a geometric pattern. This forms a truss-like structure. Space frame structures are both lightweight and rigid.

One major advantage of this type of structure is that it can span large spaces with very few support struts or columns. It can also be used to give a modern, visually appealing aesthetic. This makes it ideal for use in public spaces. For example, the roof design of the modern section of London King's Cross railway station uses a space frame structure.

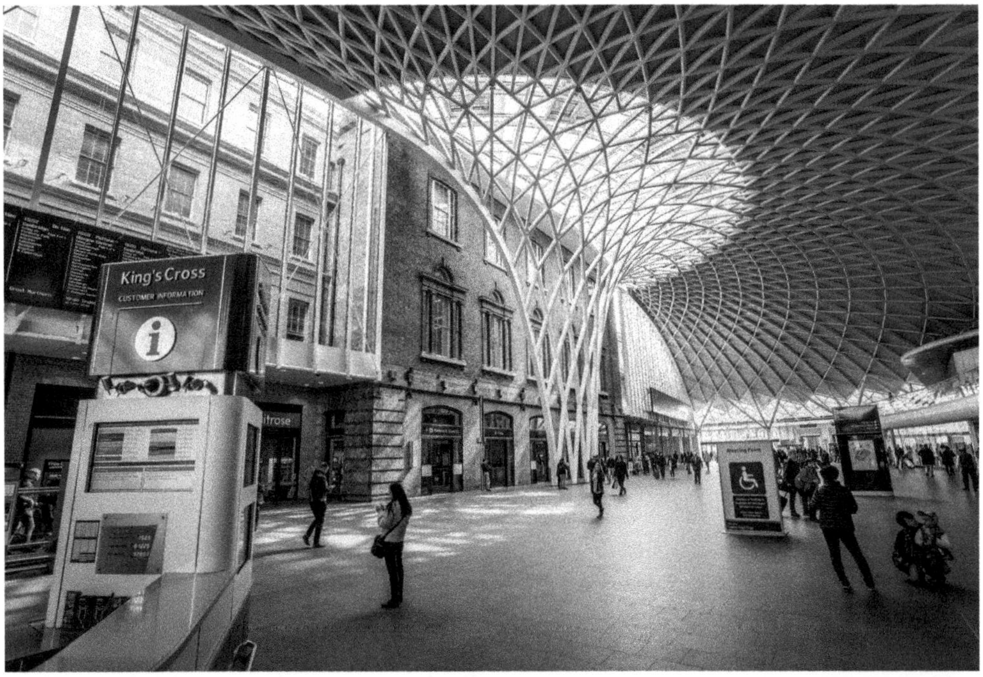

Figure 3.5.3 **Space frame structure used in the roof structure at King's Cross railway station**

Monocoque structures

A **monocoque structure** uses an external skin to support the load required. In other words, the exterior surface is also the primary structure. It does not have, or need, an internal load carrying frame. Monocoque is a French word meaning 'single shell'.

Early aircraft designs made use of this type of structure. In 1981, McClaren created the first carbon-fibre monocoque Formula 1 car.

This type of structure has the advantage of much fewer potential weak points, such as weld joins. It is also both stiff and relatively lightweight. However, it can be subject to buckling if too much compressive stress is placed on it, putting limits on its potential use.

Semi-monocoque structures were designed to overcome this problem by using both the skin and an underlying frame structure to support the load. This technique is commonly used in the design of commercial jet airliners, where the stresses that they are placed under are very high.

Figure 3.5.4 **Close-up of the roof structure at King's Cross railway station**

Figure 3.5.5 **Modern jet airliners are constructed using a semi-monocoque structure**

ACTIVITY

Investigate examples of the use of space frame and monocoque structures. Explain the advantages and disadvantages of using these types of structures in each situation.

KEY POINTS

- Static loads are constant over time.
- Dynamic loads vary over time.
- Tensile stresses can pull structural components apart.
- Compressive stresses can cause structural components to buckle.
- Space frame structures are constructed from struts that lock together in a geometric pattern.
- Monocoque structures use an external skin or shell to support the load.

Check your knowledge and understanding

1 Give an example of a dynamic load.
2 Name the stresses that cause bending in structural systems.
3 Give an advantage of using a space frame structure.

3.6 Pneumatic systems

What will I learn?

By the end of this section you should have developed knowledge and understanding of:

→ the uses of and differences between hydraulic and pneumatic systems
→ how single- and double-acting cylinders work and how their output force is calculated
→ how valves can be connected together to form simple logic circuits
→ how delays can be achieved with pneumatic systems
→ the main applications of pneumatics in engineering contexts.

Pneumatic and hydraulic systems are widely used in engineering applications ranging from simple hand tools to complex robots and production systems, for example, the systems that mass produce our favourite foods, games consoles and cars.

This section discusses the differences between pneumatic and hydraulic systems and their typical uses. It looks at how different components can be used to create circuits to achieve time delays and simple logic functions.

You will also learn about how different types of cylinders work and how to calculate their output force.

Figure 3.6.1 Valves and pipes in hydraulic machinery

Pneumatic and hydraulic circuits

Fluid-power systems use fluids to control and transmit power. This is done through the use of components such as pumps or compressors, valves, actuators and conductors.

The two types are **hydraulic systems** and **pneumatic systems**. Hydraulic systems use a liquid, such as water or oil as the control medium. Pneumatic systems use a compressible gas, such as air.

Pneumatic systems versus hydraulic systems

When deciding whether to use pneumatic or hydraulic systems, engineers must consider the speed of operation, power that can be provided and the suitability for the environment in which the system is to be used.

Pneumatic systems generally operate faster than hydraulic systems. This is because air flows through a system much more quickly and with less resistance than liquids do. In addition, liquids must be routed back to a reservoir, whereas compressed air can be put straight out into the atmosphere if needed. A reservoir is a container that stores the fluid used to supply a hydraulic system, or the air in the case of a pneumatic system.

Another major advantage of pneumatic systems is that they are clean. In hydraulic systems, faulty valves or seals can result in oil or other liquids leaking into the surrounding environment. This makes pneumatic systems far more suitable for use in food-processing applications, where avoiding contamination is very important.

However, pneumatic systems cannot produce the same levels of force as hydraulic systems. This makes hydraulic systems more suited to applications such as in heavy lifting and digging equipment.

Pneumatic systems can be more expensive than hydraulics due to the amount of energy needed to compress the air used in the system.

Common pneumatic circuits and components

Pneumatic circuits, like electrical and electronic circuits, are drawn using symbols to create a circuit diagram or a schematic. A list of common symbols for components used in pneumatic circuits can be found in the Engineering symbols section of this book (p. 218). Pneumatic circuits can be used to achieve functions such as delays and simple logic. You also need to understand how the different types of **cylinders** work.

Single and double-acting cylinders

Cylinders are pneumatic components that produce a linear or reciprocating motion. They do this by moving a piston. The two main types are: single-acting cylinders and double-acting cylinders.

Single-acting cylinders

Single-acting cylinders use pneumatic pressure to create a force or linear motion that acts in one direction only. This is usually outwards from the component itself. A spring is used for the cylinder to return to its original position. The spring can result in more force being needed to push against it, which is a disadvantage of this type.

Double-acting cylinders

Double-acting cylinders can both extend and retract using pneumatic pressure. They can therefore produce reciprocating motion. This is done using two ports. One of these is used to bring in air for the outwards motion and the other for the inwards motion.

The force output from a cylinder can be calculated using the formula:

Force (F) = Pressure (P) × Area (A)

The area is the surface area of the piston and will sometimes need to be calculated before placing the value into the formula. The area is given by:

$A = \pi r^2$

where r is the radius of the cylinder and π is 3.14 to three significant figures.

Figure 3.6.2 **A hydraulic digger**

ACTIVITY
Make a list of the advantages and disadvantages of pneumatic and hydraulic systems.

Figure 3.6.3 **A compressed air cylinder**

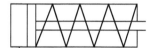
Figure 3.6.4 **Symbol for a single-acting cylinder**

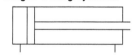
Figure 3.6.5 **Symbol for a double-acting cylinder**

KEY WORD

Cylinder: a component that uses pneumatic pressure to produce linear or reciprocating motion.

MATHEMATICAL UNDERSTANDING

E10: Calculating force
This example shows how to calculate the force output of a cylinder.
The radius of the piston is 10 mm and the pressure in the cylinder is 0.5 N mm^{-2}.
First calculate the surface area:
Area = $\pi \times 10^2$
Area = $3.14 \times 10 \times 10$
Area = 314 mm^2
Now calculate the force:
Force = Pressure × Area
Force = 0.5 × 314
Force = 157 N

Delay circuits

A **delay circuit** can be created by using a unidirectional-flow control valve and a reservoir connected in series. The valve restricts the flow of air into the reservoir. This causes the reservoir to fill up slowly, resulting in a time delay. The length of this can be increased or decreased using an adjustable control valve. In the example in Figure 3.6.6, a five-port valve is then operated when the pressure in the reservoir reaches the required level. Air is then released and the piston in the cylinder moves. Three-port valves can be used to both start the time delay and reset the system.

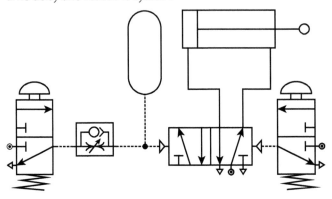

Figure 3.6.6 **A pneumatic time-delay circuit**

Logic circuits

Valves placed in series or in parallel can be used to create simple **logic circuits**.

An AND circuit consists of two valves connected in a series. The button on each of both valves must be pressed for the cylinder to operate. In other words, the cylinder will only operate when both valve A and valve B are pressed.

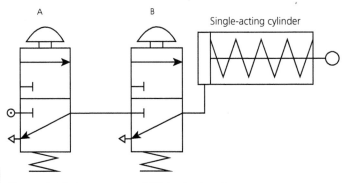

Figure 3.6.7 **A pneumatic AND logic circuit**

An OR circuit consists of two valves connected in parallel. The cylinder will operate if any of the two buttons (one on each valve) are pressed. In other words, the cylinder will operate when the button on either valve A or valve B is pressed.

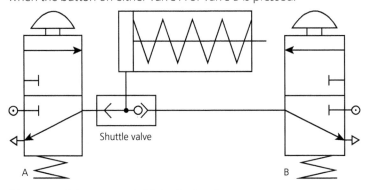

Figure 3.6.8 **A pneumatic OR logic circuit**

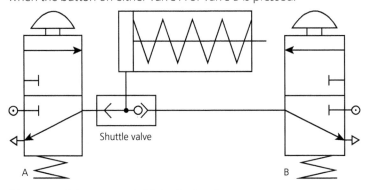

KEY WORDS

Delay circuit: a circuit that uses a unidirectional-flow valve and a reservoir connected in series to create a time delay.

Logic circuits: circuits that can create AND and OR functions through the use of valves.

Figure 3.6.9 **A pneumatically controlled robot arm**

Applications of pneumatics

The high speed, accuracy and precision of pneumatic systems makes them ideal for use in robotics applications, for example, controlling a small robot arm that must be able to make small, fast and precise movements.

Pneumatics are also heavily used in factory production lines, increasing the amount of automation in manufacturing. Assembly tools can be powered and controlled by pneumatics, thus reducing the amount of human labour needed, and increasing both speed and accuracy of production.

Tools such as drills, saws, screwdrivers and hammers can all be pneumatically assisted for greater power and speed of operation. A jackhammer, for example, is a type of pneumatic drill used in the construction industry to break up concrete, pavements and roads. These can also be hydraulically powered.

Figure 3.6.10 **A jackhammer being used in a construction project**

KEY POINTS

- Pneumatic systems use compressible air whereas hydraulic systems use liquids.
- Cylinders are pneumatic components used to produce linear or reciprocating motion.
- Valves can be connected in series or parallel to create AND or OR logic circuits.
- Pneumatics are used in production lines to increase the speed and accuracy of manufacture and assembly.

Check your knowledge and understanding

1 Describe the difference between a pneumatic and a hydraulic system.
2 Describe how single- and double-acting cylinders work.
3 State the formula for calculating the output force of a cylinder.
4 Explain how a pneumatic AND circuit works.

PRACTICE QUESTIONS: systems

1   The ohm is the unit of:
   a   current
   b   power
   c   resistance
   d   voltage.

2   What is the formula for mechanical advantage?
   a   MA = Load × Effort
   b   MA = Effort × Effort
   c $MA = \dfrac{Effort}{Load}$
   d $MA = \dfrac{Load}{Effort}$

3   A light sensor is an example of what type of sub-system?
   a   Driver
   b   Input
   c   Output
   d   Process

4   Which is the unit of force?
   a   Joule
   b   Ohm
   c   Newton
   d   Volt

5   Describe the differences between pneumatic and hydraulic systems.

6   Explain why pneumatic systems are often used instead of hydraulic systems in food production.

7   Describe the function of a resistor in an electronic circuit.

8   A two-spur gear train has a driver gear with 90 teeth and a driven gear with 45 teeth. Calculate the gear ratio of the system.

9   Describe how an OR and an AND function can be achieved using pneumatic components.

10 Describe the function of an electronic comparator circuit.

11 Name the three legs of a bipolar transistor.

12 Name two types of mechanical bearing.

13 An engineer needs to create an electronic timer function for a security system. They could do this using discrete components or a programmable microcontroller.

   Discuss the advantages and disadvantages of using a microcontroller over discrete electronic components for the timer.

4 Testing and investigation

Engineers use modelling and calculating to test design ideas, as well as prototypes and final products, to ensure that they have the properties needed. This helps to ensure that parts and components are suitable to meet the needs of the application.

This section includes the following topics:

4.1 Modelling and calculating

4.2 Testing

4.3 Aerodynamics

At the end of this section you will find practice questions relating to modelling and calculating.

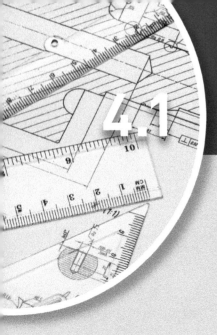

4.1 Modelling and calculating

What will I learn?

By the end of this section you should have developed knowledge and understanding of:

→ the use of simulations, modelling and calculations to predict the performance of systems

→ how electronic circuits and hydraulic/pneumatic systems are typically designed using CAD software

→ how to calculate dimensions, areas, volumes, mechanical properties and hydraulic and pneumatic forces within a system and resistance in a circuit.

Designing an engineered product requires more than just choosing which materials to use. For example, consideration must be given to the amount of material needed and you will need to ask questions such as: What size should it be? How thick should it be?

You also need to be aware that if more material is used than needed, this will increase the weight of the product and the cost. If less material is used than needed, this may mean that the product is not strong enough. Similarly, with electrical and mechanical systems (see Section 3, Systems), the components must be selected to achieve the performance needed. If the wrong components are used, the system might not work at all – or worse, they might cause safety problems.

Figure 4.1.1 Engineers need to use calculations to predict the performance of systems

Predicting performance

Users expect products to satisfy the needs that they were designed for. For example, an aircraft landing gear must support the weight of the aircraft as it lands; it could be disastrous if it collapsed the first time it was used. If someone buys a computer from a shop, they expect it to perform as expected when it is turned on.

A lot of work usually goes into the design of products to ensure that they can meet the requirements of the application they will be used for. Before a product is made, engineers normally predict and model its performance. This can be a very complex process; many products include tens, hundreds or even thousands of components or parts. Every individual part must meet certain needs and how all these parts work together must also meet the requirements of the application.

For example, an aeroplane wing may be made from more than a dozen metal beams, covered with metal body panels, all joined together with rivets and welds. The material used in every beam and body panel will need to have a certain amount of strength, and the bolts and welds will also need to have certain properties. Even if all these parts are up to the task, how they are arranged will also affect the strength of the structure.

Similarly, an electronic system may contain many different components; for the system to work, every one of these components must fulfil its own function and work with the other components. An alarm system would not be much use if, every time it was triggered, a current overload damaged its controller or did not provide enough power to operate its siren.

Figure 4.1.2 Aircraft landing gear needs to be thoroughly tested as the consequences of failure are serious

Calculating and simulating performance

Engineers use calculations to predict the performance of parts, products and systems. These can simulate and model how the product or system will perform in use. For a simple application, these calculations may be carried out manually (see below). However, for a complex product with many parts or components that interact with each other, this would be very time consuming and there is the risk that one error in calculation could result in the failure of the product or system. For this reason, computer aided design (CAD) software is often used to automate the process. The CAD software can, in a fraction of a second, carry out tens of thousands of calculations to see how all the different parts of the system interact. This can be carried out virtually (on screen), saving money as no parts or materials need to be purchased before the engineer is confident that the product or system will work. Also, different design ideas can be tested on screen quickly and easily, without waiting for new parts to be bought, which can stall the production schedule, incurring extra costs.

CAD software is frequently used to predict performance when:
- designing electronic circuits. This can range from creating systems diagrams to circuit schematics and even printed circuit board layouts (see Section 4.2)
- calculating hydraulic or pneumatic forces (see Section 4.2)
- carrying out stress analysis, such as calculating the stresses encountered in an aircraft wing or in a bridge structure.

Although CAD software can automate the simulation and modelling of design ideas, engineers do need to understand the principles and calculations upon which the software is based. This includes the ability to carry out basic calculations manually.

Calculating sizes: dimensions of a triangle

It is often necessary to calculate the lengths, widths or angles within triangular components, for example when marking out parts for machining operations.

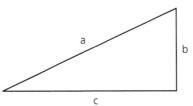
Pythagoras' theorem

If the dimensions of two sides of a right-angled triangle are known, **Pythagoras' theorem** can be used to calculate the dimensions of the third side.

Figure 4.1.3

Pythagoras' theorem states: $a^2 = b^2 + c^2$

Rearranging this formula:

$$a = \sqrt{(b^2 + c^2)}$$

$$b = \sqrt{(a^2 - c^2)}$$

$$c = \sqrt{(a^2 - b^2)}$$

Figure 4.1.4 Metal part being turned with a taper at an angle

Trigonometry

Trigonometry can be used to determine the dimensions of a right-angle triangle. This might be necessary when setting the distances to be travelled by a laser cutting head or CNC milling machine, or determining the dimensions of a taper being produced using a lathe.

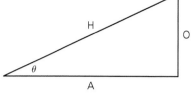

Figure 4.1.5

The angle θ is related to the hypotenuse (H), opposite (O) and adjacent (A) by the equations for the tangent (tan), sine (sin) and cosine (cos):

$$\tan\theta = \frac{O}{A}$$

$$\sin\theta = \frac{O}{H}$$

$$\cos\theta = \frac{A}{H}$$

KEY WORD

Trigonometry: the branch of maths dealing with the relationship between the sides and angles of triangles.

If at least two of the values are known, these equations can be rearranged to find unknown values. For example, rearranging to find θ:

$$\theta = \tan^{-1}\left(\frac{O}{A}\right)$$

$$\theta = \sin^{-1}\left(\frac{O}{H}\right)$$

$$\theta = \cos^{-1}\left(\frac{A}{H}\right)$$

Similarly, the equations can be rearranged to find values for A, O or H:

$$A = \frac{O}{\tan\theta} = H \times (\cos\theta)$$

$$O = A \times (\tan\theta) = H \times (\sin\theta)$$

$$H = \frac{A}{\cos\theta} = \frac{O}{\sin\theta}$$

MATHEMATICAL UNDERSTANDING

M1.6: Calculate angles of a triangle using trigonometry

Question

A piece of material needs to be marked out for cutting. Calculate the angle θ of Figure 4.1.6 below.

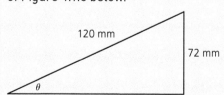

Not to scale

Figure 4.1.6

Solution

Using trigonometry, $\sin\theta = \frac{72}{120} = 0.6$

Therefore $\theta = \sin^{-1}0.6 = 36.9°$

<table>
<tr><td>

</td></tr>
</table>

KEY WORDS

Area: the size of the surface of a shape.

Volume: the amount of space that an object or substance occupies.

Density: the mass of material per unit volume.

Area, volume and density

Area, **volume** and **density** are often used to calculate the properties of components or consumable needs. For example, volume might be used with density to determine weight, which might be important if the product has to fly or float.

Area

The area of a component may be needed to calculate the amount of material needed or the amount of material that will be waste when a component is cut out.

The formulae for the areas of simple shapes are:
- Area of a rectangle = length × width = L × W
- Area of a circle = πr^2
- Area of a triangle = half (base × height) = ½ (B × H)

(To change a fraction to a decimal, you divide the top number by the bottom number, so $\frac{1}{2}$ = 1 ÷ 2 = 0.5)

Figure 4.1.7 Circular pieces cut in a metal sheet

MATHEMATICAL UNDERSTANDING

E3: Area of a circle

Question

A company is cutting a circle of radius of 0.3 m from a sheet of material that is 0.7 m square. After the circle has been cut out, the rest of the material will be waste.

Calculate the percentage of material that will be waste.

Solution

Cross-sectional area of circle = πr^2 = $\pi \times 0.3^2$ = 0.283 m²

Amount of material used = 0.7^2 = 0.49 m²

Amount of waste material = 0.49 − 0.283 = 0.207 m²

Percentage of waste = $\frac{0.207}{0.49} \times \frac{100}{1}$ = 42.2%

The areas of complex shapes can be calculated by breaking them down into simple shapes. For example, the shape in the figure below can be broken down into a rectangle and a triangle, as follows:
- The area of the rectangle = length × width = 0.8 × 0.9 = 0.72 m²
- The area of the triangle = half (base × height) = 0.5 × (0.8 × 0.3) = 0.12 m²

The total area is the sum of the shapes = 0.72 + 0.12 = 0.84 m²

Figure 4.1.9 Laser cutting a shape

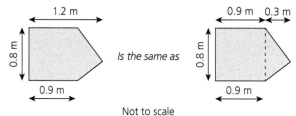

Figure 4.1.8

MATHEMATICAL UNDERSTANDING

E1: Area of a rectangle and E3: Area of a circle

Question

Calculate the area of the part shown below:

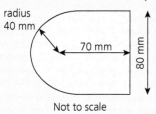

Figure 4.1.10

Solution

Area of semi-circle $= \dfrac{\pi r^2}{2} = \dfrac{\pi \times 40^2}{2} = 2513$ mm^2

Area of rectangular section $=$ length $\times$ width $= 70 \times 80 = 5600$ mm^2

Total area $= 2513 + 5600 = 8113$ mm^2

Volume and density

Volume may need to be calculated to determine:
- the capacity of a container of coolant
- the amount of material needed to make a product.

The formulae for simple shapes are:
- Volume of a cuboid, V $=$ length $\times$ width $\times$ height $=$ L $\times$ W $\times$ H
- Volume of a cylinder, V $=$ area of circle $\times$ length $=$ A $\times$ L $= \pi r^2 \times$ L

The volume can also be used with the density to determine the mass of the material.
The formula for the density is:

$$\text{Density, } \rho = \frac{\text{mass}}{\text{volume}} = \frac{m}{V}$$

This can be rearranged to give the mass:

$$\text{Mass, } m = \rho \times V$$

MATHEMATICAL UNDERSTANDING

E4: Volume of a cylinder

Question

A container for polymer resin is cylindrical in shape, with internal dimensions of radius 0.2 m and height 0.9 m. The density of the resin is 1800 kg m^{-3}. Calculate the mass of the resin that the container can hold when full.

Solution

Volume of the container $= \pi r^2 \times$ L $= 0.113$ m^3

$$\text{Mass} = \rho \times V = 1800 \times 0.113 = 203.4 \text{ kg}$$

Figure 4.1.11 Plastic containers for machine oil

Conversion of load/extension to stress/strain

Products may have to be able to resist a certain load without a significant change in dimensions–for example, the structural members of a bridge must support the load passing over the bridge without sagging. As described in Section 1.1 Materials and their properties, when the strength of a material is tested, the testing machine records the force or load applied and the change in length of the test piece. These values need to be converted into the **stress** in the material and the **strain**. The formulae for these are:

- Stress, $\sigma = \dfrac{\text{force}}{\text{cross-sectional area}} = \dfrac{F}{A}$

- Strain $= \dfrac{\text{change in length}}{\text{length}} = \dfrac{\delta l}{l}$

These equations are also used to calculate the mechanical properties of parts in service. It should be remembered that strain does not have units.

MATHEMATICAL UNDERSTANDING

E7: Stress

Question

A tensile test was carried out on a test piece with a radius 8 mm. The force applied when the material started to yield was 25 326 newtons. Calculate the yield stress of the material.

Solution

Given force F = 25 326 N and cross-sectional area $A = \pi r^2 = \pi \times 8^2 = 201\,\text{mm}^2$

$$\text{Stress } \sigma = \frac{F}{A} = \frac{25\,326}{201} = 126\ \text{Nmm}^{-2}$$

MATHEMATICAL UNDERSTANDING

E8: Strain

Question

A load attached to the end of a 2.5 m metal bar causes the bar to extend to a length of 2.55 m. Calculate the strain in the bar due to the load.

Solution

$$\text{Strain, } \varepsilon = \frac{\text{change in length}}{\text{original length}}$$

$$\varepsilon = \frac{0.05}{2.5} = 0.02 \left(\text{or } 2 \times 10^{-2}\right)$$

Young's Modulus and stiffness

Young's Modulus is a property of a material that indicates its **elasticity** and **stiffness**. It is calculated using the formula:

$$\text{Young's Modulus } E = \frac{\text{stress}}{\text{strain}} = \frac{\sigma}{\varepsilon}$$

The stress and strain are read from the stress/strain chart for the material. The value of the stress used must be equal to (or below) the **yield stress** of the material. If the value taken is below the yield stress, this does not change the value calculated for Young's Modulus, as the corresponding strain will be lower.

When the Young's Modulus of a material is known, rearranging the equation allows calculation of either the stress or the strain that will be experienced.

MATHEMATICAL UNDERSTANDING

E9: Young's Modulus

Question

A round bar of material is 200 mm long. When subjected to a stress of 900 N mm^{-2}, the length of the material increased to 203 mm. The material returned to its original length when the stress was removed.

Calculate the Young's Modulus of the material.

Solution

$$\text{Strain, } \varepsilon = \frac{\text{change in length}}{\text{original length}} = \frac{(203-200)}{200} = 0.015 \left(\text{or } 1.5 \times 10^{-2}\right)$$

$$\text{Young's Modulus } E = \frac{\text{stress}}{\text{strain}} = \frac{\sigma}{\varepsilon} = \frac{900}{0.015} = 60 \text{ kN mm}^{-2}$$

KEY WORDS

Young's Modulus: a measure of elasticity, equal to the stress acting on the material divided by the strain.

Elasticity: the ability of a material to return to its original shape when the load upon it is removed.

Stiffness: the rigidity of an object.

Yield stress: the stress at which a material starts to permanently deform.

Factors of safety

For some products, there can be serious safety issues if the product breaks in use (for example, aircraft components, gas cylinders, pressure vessels, bridges and lifting equipment). In these cases, engineers will often apply a **factor of safety**. This is the ratio of the yield stress of the material to the maximum load that may be applied to it. It has no units.

Figure 4.1.12 Gas cylinders should be tested by applying a factor of safety

In practice, this means that a product is potentially much stronger than is actually needed for the design. The reason for this is to give confidence that the product will not fail, for example due to:
- flaws or defects in the material
- manufacturing errors
- corrosion of the product
- possible misuse of the product.

The formula for the factor of safety is:

$$\text{FoS} = \frac{\text{yield stress}}{\text{load}} = \frac{\sigma_y}{L}$$

MATHEMATICAL UNDERSTANDING

E11: Factor of safety

Question

An engineer is designing a gas cylinder that will have an internal pressure of up to 45 kN mm^{-2}. If the factor of safety is 3, what is the minimum value of the yield strength for the material that would be acceptable?

a 15 kN mm^{-2}

b 45 kN mm^{-2}

c 135 kN mm^{-2}

d 180 kN mm^{-2}

Solution

a 135 kN mm^{-2}

Rearranging

$$\text{FoS} = \frac{\sigma_y}{L}$$

$$\sigma_y = \text{FoS} \times L = 3 \times 45 = 135 \text{ kN mm}^{-2}$$

Forces within/applied to a component or a system

The performance of the various types of system detailed in Section 3 can also be calculated.

Mechanical advantage

The **mechanical advantage** represents how much easier or harder it is for a mechanism to carry out an action. For example, a pulley system can be designed to raise a load many times larger than the force applied to lift it (see Section 3.2). Mechanical advantage has no units and is calculated using the formula:

$$\text{Mechanical advantage } MA = \frac{\text{load}}{\text{effort}} = \frac{F_b}{F_a}$$

> **MATHEMATICAL UNDERSTANDING**
>
> **E15: Mechanical advantage**
>
> A pulley system is being used to lift a sack of components with a mass of 450 N. The mechanical advantage of the system is 6. Calculate the force that must be applied to raise the load.
>
> Rearranging
>
> $$MA = \frac{F_b}{F_a}$$
>
> $$F_a = \frac{F_b}{MA} = \frac{450}{6} = 75 \text{ N}$$

Gear ratio

The **gear ratio** determines the speed of rotation of two (or more) gears in contact by their teeth. The first gear, providing the input into the system, is referred to as the driver gear. The output gear is referred to as the driven gear.

The gear ratio should always be presented as two numbers (whole numbers) in the lowest possible integer values, for example, 5:1, 3:8. If the first number is higher, this represents a gear train where the speed is being reduced; if the second number is higher, this represents a gear train where the speed is being increased. The formula to calculate the gear ratio is:

Gear ratio = number of teeth on driven gear / number of teeth on driver gear = N_{driven} / N_{driver}

As noted in Section 3.2 Mechanical systems, the amount of torque transmitted is the inverse of the gear ratio; i.e. the higher the increase in speed, the lower the force transmitted through the gear.

The speed of the gears is in proportion to the gear ratio, i.e. Speed_{driver} / $\text{Speed}_{driven} = N_{driven}$ / N_{driver}

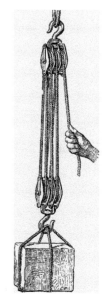

Figure 4.1.13 Pulleys have been used for hundreds of years – this picture is from 1895!

For example, if the driver gear rotates at 120 revolutions per minute (rpm) and the gear ratio is 3:1, then the driven gear will rotate at 40 rpm.

Figure 4.1.14 Interlocking metal gears

MATHEMATICAL UNDERSTANDING

E14: Gear ratio

Question

In a simple gear train comprising of two gears, the number of teeth on the driver gear is 81 and the number of teeth on the driven gear is 18. The driver gear is rotating at a rate of 86 rpm.

a   What is the gear ratio of the system?

b   What is the speed of rotation of the driven gear?

Solution

a   Gear ratio = 18:81 = 2:9

b   Rearranging

$$\frac{\text{Speed}_{driver}}{\text{Speed}_{driven}} = \frac{N_{driven}}{N_{driver}}$$

$$\text{Speed}_{driven} = \text{Speed}_{driver} \times \left(\frac{N_{driver}}{N_{driven}}\right)$$

$$\text{Speed}_{driven} = \frac{86 \times 9}{2} = 387 \text{ rpm}$$

KEY WORD 🔑

Velocity ratio: the ratio of the distance through which any part of a machine moves, to that which the driving part moves during the same time.

Velocity ratio

Velocity ratio is similar to gear ratio, except that instead of the number of teeth, the ratio is normally between the dimensions of wheels, for example in a pulley. As for the gear ratio, the velocity ratio should be presented as two numbers in the lowest possible integer values, for example: 1:4, 9:5. The formula for the velocity ratio is:

$$\text{Velocity ratio} = \frac{\text{size of output wheel (or pulley)}}{\text{size of input wheel (or pulley)}}$$

The same value of velocity ratio is calculated when using either the radius, diameter or circumference, provided that the same type is compared for both wheels; for example, radius:radius, diameter:diameter or circumference:circumference.

Designing pneumatic and hydraulic systems: pressure

Pressure calculations may be used to calculate the output force of a pneumatic or hydraulic system, or the force applied to a material by a press. The formula for pressure is:

$$\text{Pressure} = \frac{\text{Force}}{\text{Area}} = \frac{F}{A}$$

The SI unit for pressure is the pascal (Pa), which is equal to one newton per square metre.

Figure 4.1.15 Air cylinder and pneumatic valve

Designing electrical circuits: resistance

Electrical systems were described in Section 3.3. Ohm's Law relates the current, voltage and resistance in an electrical circuit. It can be presented as a triangle to assist recall:

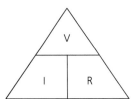

Figure 4.1.16 Ohm's Law

Using Ohm's Law:

$$V = IR$$

$$I = \frac{V}{R}$$

$$R = \frac{V}{I}$$

The resistance of a group of resistors can be calculated depending upon how they are arranged.

The resistance of resistors in series, Figure 4.1.17, is given by the sum of the resistor values:

$$R_T = R_1 + R_2$$

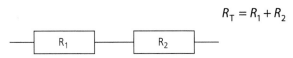

Figure 4.1.17

MATHEMATICAL UNDERSTANDING

E12: Ohm's law

A circuit contains the following arrangement of resistors:

22 Ω	68 Ω

Figure 4.1.18

The current flowing in the circuit is 2 A.
a Calculate the total resistance.
b The current flowing in the circuit is 20 mA. Calculate the voltage.

Solution

a $R_T = R_1 + R_2$
   $= 22 + 68$
   $= 90 \, \Omega$

b Using Ohm's Law

   $V = I \times R$
   $= 0.02 \times 90$
   $= 1.8$ volts

Cost

Cost can refer to:
- the sum of the components purchased to make the product
- the cost of labour to machine or assemble the product: labour cost = time × charge rate
- the total manufacturing cost of a product: total cost = parts + materials + cost of labour

MATHEMATICAL UNDERSTANDING

M13: Perform calculations using time and cost

Question

The cost of the components needed to make a product is shown in the table below.

Table 4.1.1 **Cost of components**

Component:	Cost each, £:	Quantity needed:	Total cost
Base	0.48	1	
Sides	0.32	3	
Cover	0.65	1	
Screws	0.05	4	
		Total:	

The labour cost to make the product is £2.21.

a Complete the table to show the total cost of materials in the product.
b Calculate the total cost of making the product.

Solution

a

Table 4.1.2 **Cost of components: answer**

Component:	Cost each, £:	Quantity needed:	Total cost
Base	0.48	1	0.48
Sides	0.32	3	0.96
Cover	0.65	1	0.65
Screws	0.05	4	0.20
		Total:	2.29

b £4.50

Designing electrical circuits: resistance

Figure 4.1.19 Using graphs

Graphs

In addition to carrying out calculations, engineers use graphs to predict performance. Many types of graphs may be used, including bar charts, line graphs (see example in Section 1.3), pictograms and pie charts. Using graphs includes:

- reading and interpreting data from graphs
- plotting graphs from provided data
- drawing an appropriate trend line onto plotted data
- determining the slope of a graph.

The formula for a straight line on a line graph is:

$$y = mx + c$$

where y is the value on the y axis, c is the value of y where the line crosses the y axis (i.e. where $x = 0$), and m is the gradient or slope. M can be calculated using x and y co-ordinates of the line from any two points on the line, although typically one will be near the base of the line and the other towards the end:

$$m = (x_{high} - x_{low}) / (y_{high} - y_{low})$$

ACTIVITY

Using the internet, find examples of four different types of graph that are used in manufacturing companies. Why is each type of graph used for its purpose, rather than one of the other types?

MATHEMATICAL UNDERSTANDING

M4.1: Translate information between graphical and numeric form

Question

A company that makes machine tools has collected data on which industry sectors purchase their products. The results are presented in a pie chart. The total value of the sales is £24 million.

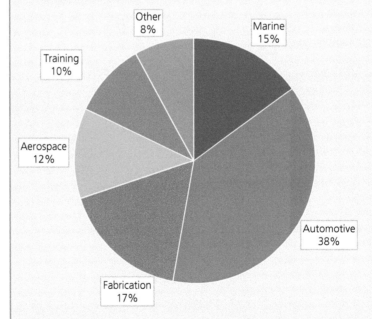

Figure 4.1.20

a Calculate the value of the automotive segment.
b Calculate the fraction of the total market value that is for marine. Give your answer as a fraction in its lowest form.

Solution

a $24\,000\,000 \times 38/100 = £9\,120\,000$ (£9.12 million)
b $15/100 = 3/20$

M4.2: Plot two variables from experimental or other data

Question

A company asked 60 customers to identify the most important features of their product. The responses are shown in the table below.

Table 4.1.3 **Customer feedback**

Response from customer:	Number of customers:	Percentage of total:
Cost	24	40
Speed of delivery	15	
Colour		15
Well packaged	8	13
No preference	4	7
Total	**60**	**100**

a   Insert the missing values in the table.
b   Use the information in the table to create a bar chart showing the percentage of customers who gave each response.

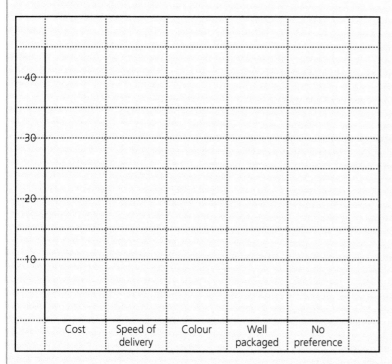

Figure 4.1.21

Solution

a Speed of delivery: 25%, colour: 9 customers

b The graph should look like:

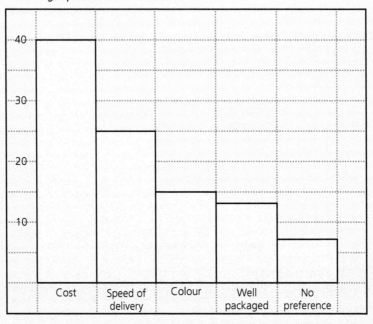

Figure 4.1.22

MATHEMATICAL UNDERSTANDING

M4.4: Determine the slope of a graph

Question

The graph below shows data an engineering company has collected data on how long cutting tools last before breaking on one of their machines.

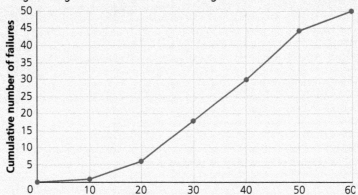

Calculate the slope of the graph.

Solution

$$m = \frac{(x_{high} - x_{low})}{(y_{high} - y_{low})}$$

Taking two points on the line:

Where x = 10; y= 1; and

Where x = 60; y = 50

$$m = \frac{60-10}{50-1} = \frac{50}{49} = 1.02$$

Check your knowledge and understanding

1 A piece of material of volume 510 cm^3 has a mass of 1.224 kg.
  Calculate the density of the material.

2 A round-section part of radius 6 mm is supporting a load of 450 kN.
  Calculate the stress in the part.

3 At its yield strength of 100 MPa, a material has a strain of 8×10^{-4}.
  Calculate the Young's Modulus of the material.

4 A metal of yield strength 135 MPa is being used to make a component that must resist a stress of 45 MPa.
  Calculate the factor of safety.

5 Three resistors are in series: 330 ohms, 470 ohms and 240 ohms.
  Calculate the total resistance.

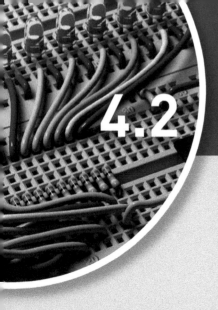

4.2 Testing

What will I learn?

By the end of this section you should have developed knowledge and understanding of:

→ how tests can be conducted to determine the tensile and compressive strength of an object

→ how circuits can be simulated using computer aided design (CAD) software and the advantages of doing this

→ how circuits can be modelled using physical components and the advantages of doing this

→ how hydraulic and pneumatic forces are calculated

→ the differences between destructive and non-destructive testing

→ how programs such as a microcontroller can be modelled, simulated and changed to improve performance

→ the application of quality control checks including the use of tolerances.

An engineered product or system is little use to anyone if it doesn't work correctly. A security system could allow burglars to enter a building if the sensors are not working correctly, or a bridge could collapse if the materials used are not structurally sound.

This chapter looks at the different ways that materials and systems are tested. You will learn about how materials can be checked for compressive and tensile strength. You will also learn about the different ways that electronic and programmable systems are modelled and prototyped.

In addition, this chapter explains the difference between destructive and non-destructive testing and how quality control methods are used to ensure successful engineering outcomes.

KEY WORDS

Tensile strength test: a test in which a material or object is subjected to tension until it fails.

Compressive strength test: a test to determine the maximum compressive load that an object can take before failing.

Ultimate tensile strength (UTS): the maximum amount of tensile force that an object can take before it fails or breaks.

Methods of testing and evaluating materials and structural behaviour under load

Engineers must be able to test how materials and structures behave when different forces and stresses are acting on them. This is so they can understand how suitable the materials are for the application that they are to be used in.

Two of the most important tests that engineers make use of are **tensile strength tests** and **compressive strength tests**. Tensile stresses attempt to stretch or pull a material apart, whereas compressive stresses attempt to squash a material. More information on different types of structures and the forces that act on them can be found in Section 3.5, Structural systems.

Tensile strength test

A tensile strength test subjects an object to pulling forces until it fails. This gives the **ultimate tensile strength (UTS)**, which is the maximum amount of tension that the object can take before it fails or breaks. The most common piece of equipment used for this purpose is a universal testing machine. This can also be used to measure compressive strength.

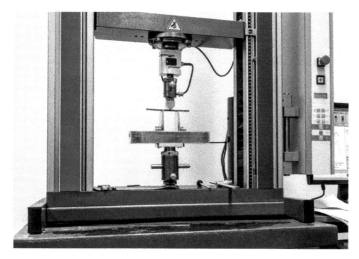

Figure 4.2.1 **A machine for testing the tensile strength of a material**

Compressive strength test

A compressive strength test subjects an object to pushing forces until it fails. This gives the maximum compressive load that the object can cope with before it fractures. A suitable compression testing machine is used. The object is placed between two plates and a load gradually applied. The results are then recorded.

ACTIVITY
Select a range of materials that you have studied. Perform appropriate tests to determine the tensile and compressive strengths of each material.

Modelling and simulating electronic circuits

Simulating and modelling circuits allows engineers to understand how they will work. It also allows engineers to correct any errors identified, before the circuits are made using permanent construction methods. This can save a great deal of time and money.

For example, an error in the capacitor value could result in a timer circuit that produces the wrong length of delay. If a batch of this circuit was produced, the manufacturing company would have to remake them all and deadlines for delivery to the customer would be missed. If the circuit had been modelled or simulated previously, this error would have been found and corrected before the batch was produced.

Circuits can be simulated virtually using computer aided design (CAD) software, or modelled physically using real components.

CAD simulations

CAD software allows circuits to be drawn and tested on screen. **CAD circuit simulations** reduce costs and increase speed and efficiency, as physical components only need to be ordered when the simulation has indicated that the circuit design will provide the required function. Changes can be made to the design quickly and easily with just a few clicks of a mouse.

Some CAD packages allow engineers to go through the complete design process, from a systems diagram to a circuit schematic and, if required, a printed circuit board layout. They can test the design at each stage to identify any problems. Some even have built-in software for programming and simulating the operation of microcontrollers (covered in more detail later in this section).

One potential problem with CAD simulations is that the components do not always act in the same way as real components in certain situations or circumstances. As a result, it is usually a good idea to model a circuit with actual components as well as simulating it virtually. Access to this software and hardware can also add to costs.

KEY WORD

CAD circuit simulation: a simulation of how a circuit works produced using computer software.

Figure 4.2.2 **Circuits can be simulated using CAD software**

Physical modelling

There are several different ways of modelling a circuit using physical components. This can take more time than producing a computer simulation but will produce very accurate results.

One common way of producing a **physical circuit model** is using prototyping board, sometimes known as 'breadboard'. This is a plastic board with holes for placing the components into. Underneath the plastic are rows of metal strips that make electrical connections between the holes. Components and wires are slotted into the holes, thus forming a circuit.

The main advantage of breadboards is that they are completely solderless. This means that components can be put in, pulled out and replaced very quickly and with no damage to them. However, when building complex models, breadboards can get very large and the layout can become difficult to follow. As a physical circuit is being produced they require storage space which can impact on costs.

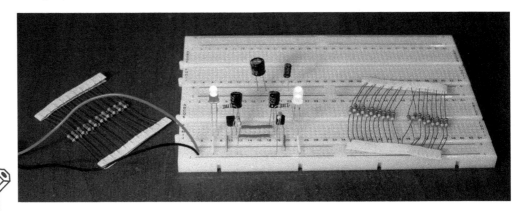

Figure 4.2.3 A breadboard being used to model a simple circuit

An alternative to breadboards is stripboard. This is a plastic board with rows of copper strips on top and holes for the placement of components. In contrast to using breadboard, the components must be physically soldered to the strips to make a circuit. This makes changing the circuit more difficult and time consuming. Many components cannot be re-used after soldering and replacements must be purchased for new versions of the model. Stripboards do however produce a sturdier, more robust outcome, and are sometimes used to produce final, permanent circuits.

The specific types of circuits that you need to be able to simulate and model using CAD software and physical methods are shown in Sections 3.3 Electrical systems and 3.4 Electronic systems.

Calculating hydraulic/pneumatic forces

Engineers also need to be able to use modelling and simulation methods to calculate hydraulic and pneumatic forces (see Section 3). CAD software can be used to achieve this or they can be performed manually. For example, when designing a pneumatic system, it could be necessary to know the output force of a cylinder to ensure that it will work as expected. If this force is too large or small then the cylinder could be replaced with one that produces the required results.

The output force of a cylinder in a fluid power system is given by the formula:

Output force = **Pressure** inside the cylinder × Area of the piston

Calculating output forces in hydraulic/pneumatic systems is covered in more depth in Section 3.6 Pneumatic systems (page 103).

Destructive and non-destructive testing

Testing of an engineered product can be either **destructive** or **non-destructive**.

Destructive testing is where the product is tested to the point where it is damaged or destroyed. The main purpose is to find out where its failure point lies. It also enables engineers to find weaknesses that might not be obvious during the normal day-to-day use of the product. The main disadvantage is that the object being tested cannot be used again, which can impact on costs. An example is the crash testing of cars to see how they react to high speed collisions with each other.

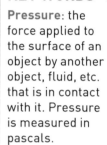

KEY WORDS

Pressure: the force applied to the surface of an object by another object, fluid, etc. that is in contact with it. Pressure is measured in pascals.

Destructive testing: testing that results in the damage or destruction of the object being tested.

Non-destructive testing: testing that does not result in damage to the object being tested.

Figure 4.2.4 Crash testing is a form of destructive testing

Non-destructive testing does not cause the product damage. Use of this type of testing can save time and it can be employed at different stages of the production process. An example of non-destructive testing is visually checking the surface finish of a product to ensure that it looks smooth enough. An industrial example is the use of ultrasonic testing of the axles on trains.

Testing control programs for programmable devices through modelling and enactment

It is not only materials and circuits that need to be tested; programmable devices must also be checked to ensure that they are working as expected. This can be done in two ways – through virtual (on screen) modelling or by physically downloading the program onto the circuit to be controlled by the programmable device.

Virtual modelling

Programs for programmable devices, such as microcontrollers, are written on screen using a dedicated programming language or software. For example, flowchart software or block-based editors are commonly used in school-based projects. More information about the ways that microcontrollers are programmed can be found in Section 3.4 Electronic systems.

Almost all modern programming software allows the writer of the program to check how the program works on screen. Some software even contains a built-in simulation of the circuit itself so the user can see the program running and its effect on the circuit.

The advantage of this is that the program can be tested before the physical circuit is required, or even assembled. This means that any required changes to the program can be made quickly and easily. Sometimes issues with how a programmable circuit performs are due to poor assembly or errors in the circuit design itself. Virtually testing and refining the program therefore aids fault finding as the program can be ruled out as a source of problems if the finished system does not work as expected. Potential alternatives to how the system could operate can also be modelled.

The simulation process usually involves a 'run through' of the program, with a display showing the results of this to the user. This could be in the form of a software dialogue box, for example, showing when the output devices are on (1) or off (0). This is usually done in real time (live, or as it happens). Any parts of the program that do not make sense to the simulation software are also flagged and reported to the user. For example, on flowchart programming software, a decision box requires both a 'yes' and a 'no' response. If one of these is missing it would lead to an error being reported.

Figure 4.2.5 An engineer testing a program on screen

Physical modelling and enactment

Once a program has been written and modelled virtually, it must be downloaded onto the programmable device itself. Many modern microcontroller systems support **in-circuit programming**. This means the device does not need to be removed from the circuit for it to be programmed. Programming requires a download cable, such as USB or serial lead, that connects the computer that the program has been written on to the circuit itself. Once the program has been downloaded, this can usually be removed. Different programming systems do this in slightly different ways so you will need to check the instructions for the one that you are using.

A series of tests can then be performed to check that the programmed circuit is working as expected. For example, if a system is designed to count 1–10 each time a switch is pressed, the user would go through the process of physically pressing the switch and checking the output devices display the counting as required.

If the programmed circuit does not work correctly, then the first things to check are if it has been downloaded correctly and if there is power to the circuit. The programming software will usually inform the user if there was a download issue. If this is not the problem, then further checks would need to be made to the program and/or the circuit. If the circuit has been successfully simulated, then the issue would more likely be with the circuit itself, which will need further investigation. For example, check all the solder joints are accurate, whether there are any loose wires, and that the circuit layout is correct.

Modifying a program to improve performance

Once a microcontroller program has been written and tested, it can be modified to improve its performance. Two examples of this are adjusting motor speeds and changing **sensor threshold ranges**.

Adjusting motor speeds

Motors can be set by a program to rotate in a certain direction and at a certain speed. This can be changed by altering the program. In the flowchart example below, a dialogue box opens when the 'motor' command is clicked on. The speed and direction of the motor can be adjusted by moving the appropriate slide bar up and down. This can be seen operating on screen before the program is downloaded to the microcontroller. In text-based programming it would be achieved by changing the appropriate value.

> **KEY WORD**
>
> **Sensor threshold range**: the sensor range between which an action takes place in a program.

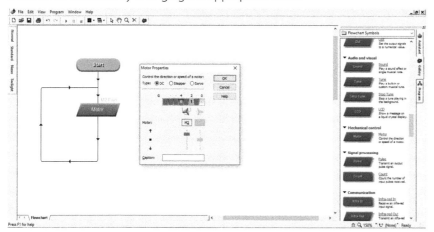

Figure 4.2.6 Changing the speed and direction of a motor using flowchart software

This could be useful when designing an electric vehicle that needs its wheels to turn at different speeds at different times.

Changing sensor threshold ranges

Analogue sensors can detect an almost infinite number of values within the limits of the device. Often, an engineer will only want an output device to do something when the sensor detects readings within (or outside) a certain range. This is called the threshold range. More information on different types of sensors can be found in Section 3.4 Electronic systems.

For example, an engineer might require a system to turn on a lamp when the light level detected by an LDR is between 10 and 100 lux. Lux is the SI unit of light. If they decided to reduce that range so that the lamp only lights between 10 and 50 lux, they would need to change the appropriate line in the program. As with the motor example above, this would be done via a dialogue box that opens when clicking on the 'analogue sensor' command in flowchart software.

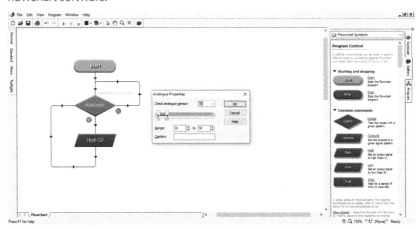

Figure 4.2.7 Changing the threshold range of an analogue sensor

> ### ACTIVITY
> Write a program to turn on an output device between a set range detected by an analogue sensor. Simulate the effect of using different threshold ranges in the program.

Quality control methods

KEY WORDS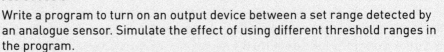

Quality control: a system designed to ensure that an engineered product meets a set of quality criteria and/or the needs of the end user.

Tolerance: the permissible limits in variation of a dimension or measured value.

Quality control (QC) is an essential part of producing engineered products. It involves checking a product after a process to ensure it meets the required quality standards.

Some examples of QC checks include:
- visual checks to ensure that the product looks as it should
- measurements of length, width, depth, etc. to ensure that the product has the correct dimensions
- weight checks to ensure that it is of the correct weight
- fitting/assembly checks to ensure that parts fit together as required.

One important part of the QC process is the application of **tolerances**. Tolerances give the permissible limits in variation of a dimension or measured value.

For example, the length of a metal bar might need to be cut to 20 mm. If the tolerance is ±0.2 mm, then anything outside of the range of 19.8–20.2 mm would be outside of tolerance.

Failure to use or adhere to tolerances can lead to parts of products that do not fit together properly. This would then require the parts to be remade, which would cost time and money.

Tolerances are usually shown on engineering drawings.

Quality control is revisited in Section 6.9.

- The tensile and compressive strengths of materials can be tested to find out the forces that they can take before failing.
- Electronic circuits can be simulated using CAD software, or modelled physically to find out how they will work and/or correct errors.
- Destructive testing results in damage to the object being tested, whereas non-destructive testing does not cause damage.
- Microcontroller programs can be tested using a computer program (simulation), or by downloading the program to the IC and running it in the circuit.
- Quality control involves checking a product after a process to ensure that it meets the required quality standards.

Check your knowledge and understanding

1 State the purpose of a tensile strength test.
2 Give two methods of modelling a physical circuit.
3 Describe the difference between destructive and non-destructive testing.
4 Explain one benefit of simulating a microcontroller program virtually.
5 Give one problem that can occur if tolerances are not adhered to when producing engineered parts for assembly.

4.3 Aerodynamics

What will I learn?

By the end of this section you should have developed knowledge and understanding of:

→ the term 'aerodynamics'
→ what is meant by 'lift', 'drag' and 'thrust' in aerodynamics
→ how aerodynamics can be applied in common engineering contexts, such as aircraft and racing vehicles.

Have you ever wondered how an aircraft can fly through the skies, or why altering the shape of a racing car can help it to go faster? Aerodynamics is fundamental to how this occurs.

This section describes what is meant by aerodynamics. It explains the terms 'lift', 'drag' and 'thrust' and their effects. You will also learn how each of these can be applied in common engineering contexts.

Aerodynamics

Aerodynamics is all about how air moves around objects, or the way objects move through the air. A common example of an engineered product that reacts to aerodynamics is an aircraft. Racing cars are also designed to be aerodynamic so they travel faster.

To understand how aerodynamics works, you must first understand the key terms of **lift**, **drag** and **thrust**. The diagram below shows how these forces act on an aircraft wing.

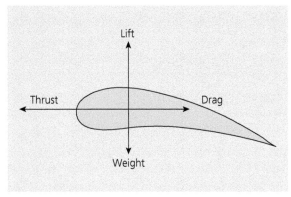

Figure 4.3.1 The forces acting on an aircraft wing

Lift

Lift is the pushing force that causes something to move upwards. It opposes (works against) the weight of the object being lifted so the lift force must exceed the weight of the object for it to fly.

In the case of an aircraft, the wings are designed to produce the lift force required for flight to take place. They have a streamlined shape with a curve on the top and a flatter bottom.

This shape allows air to flow faster over the top of the wing than it does underneath it. This results in more pressure underneath the wing than above it, causing it to move upwards.

An example of the opposite is the aerodynamics on a racing car, which are there to ensure it stays on the ground. In this case, the aerofoil section is inverted to increase downward force, thus improving grip.

Drag

The shape of an aircraft wing is also designed to reduce drag. This is the force that opposes the forward motion of an object through the air. It is often thought of as aerodynamic friction. The amount of drag is proportional to the amount of air hitting the surface of the moving object.

Drag can usually be reduced by using rounder and more narrow shapes. This not only explains the shape of wings, but also of fast racing vehicles, such as Formula 1 cars and dragsters.

Thrust

Thrust is the pushing force that causes an object to move forwards. It therefore opposes drag, so the thrust force must be greater than the drag force for forwards movement to occur. On modern commercial airliners, the thrust is provided by high-powered jet engines. On a dragster, it can be provided by an internal combustion engine. A model of a drag racing car made in school might use a compressed air cylinder instead.

Figure 4.3.2 A drag racer uses an internal combustion engine to produce thrust

KEY POINTS
- Aerodynamics describes the way objects move through the air.
- Lift is the pushing force that causes an object to move upwards.
- Drag is the force that opposes the forward motion of an object through the air.
- Thrust is the pushing force that causes an object to move forwards through the air.

Check your knowledge and understanding

1 Explain what is meant by the term 'lift'.
2 State what drag opposes.
3 Give an example of an application of aerodynamic thrust?

1 A product is made using a single machining operation and a machinist can manufacture five products in 2 hours. The cost of the material used to make the product is £8. The cost of labour per hour is £30. The finished product is sold for £25.

Calculate the cost of the labour in each product.

Calculate the profit on the product as a percentage of the sales price.

2 A piece of material needs to be marked out for cutting. Calculate the length A on the figure below.

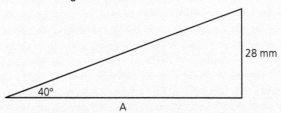

Figure not to scale

3 The internal dimensions of a rectangular oil can are 250 mm × 150 mm × 0.4 m. Calculate the volume of oil that can be held in the can.

4 The output from a motor is turning at 1500 rpm. The motor is attached to a simple gear train, comprising of two gears. The number of teeth on the driver gear is 72 and the number of teeth on the driven gear is 27.
   i   Determine the gear ratio of the gear train.
   ii Calculate the speed of the driven gear (in rpm).

5 A steel bar is subjected to a tensile stress. The Young's Modulus of the steel is 200 GPa. The length of the bar when no stress is applied is 3 m. Calculate the extension of the bar when a stress of 140 MPa is applied.

6 An engineer is designing a part for a lifting gear. The part has a square cross-section, with each side 20 mm, and will be subjected to a tensile load of 24 kN. If the design calls for a factor of safety of 3, determine the minimum acceptable yield strength of the material that will be used to make the part.

7 The voltage measured across an electrical component was 6.6 volts. The current measured flowing through the component was 30 milliamps. Calculate the resistance of the component.

8 Testing that results in damage to an engineered product is called:
   a Descriptive testing
   b Destructive testing
   c Non-destructive testing
   d Visual testing

9 QC stands for:
   a Quality control
   b Quality criteria
   c Quantity control
   d Quantity criteria

10 Explain **one** advantage and **one** disadvantage of using CAD software to model electronic circuits.

11 Define the term 'tolerance'.

12 Explain the difference between lift, drag and thrust in aerodynamics. Use a labelled diagram to aid your answer.

13 Give **two** examples of quality control checks in engineering.

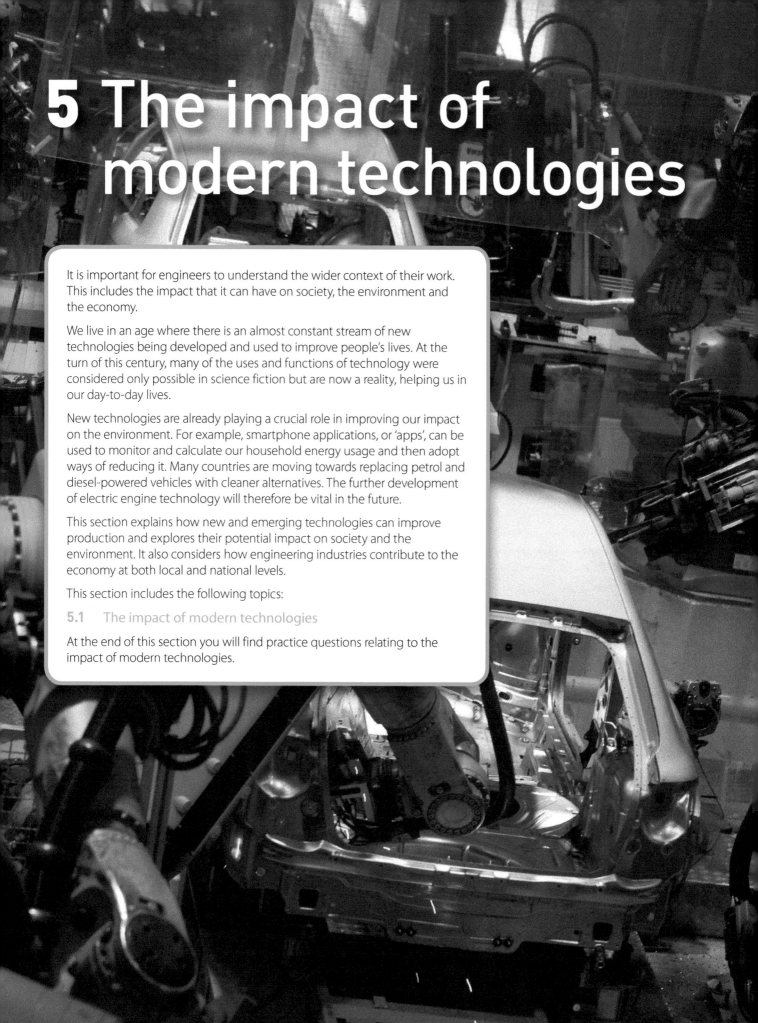

5 The impact of modern technologies

It is important for engineers to understand the wider context of their work. This includes the impact that it can have on society, the environment and the economy.

We live in an age where there is an almost constant stream of new technologies being developed and used to improve people's lives. At the turn of this century, many of the uses and functions of technology were considered only possible in science fiction but are now a reality, helping us in our day-to-day lives.

New technologies are already playing a crucial role in improving our impact on the environment. For example, smartphone applications, or 'apps', can be used to monitor and calculate our household energy usage and then adopt ways of reducing it. Many countries are moving towards replacing petrol and diesel-powered vehicles with cleaner alternatives. The further development of electric engine technology will therefore be vital in the future.

This section explains how new and emerging technologies can improve production and explores their potential impact on society and the environment. It also considers how engineering industries contribute to the economy at both local and national levels.

This section includes the following topics:

5.1   The impact of modern technologies

At the end of this section you will find practice questions relating to the impact of modern technologies.

5.1

The impact of modern technologies

What will I learn?

By the end of this section you should have developed knowledge and understanding of:

→ what is meant by a new and emerging technology
→ the impact of new and emerging technologies on production, society and the environment
→ the impact of engineering industries on society and the economy.

Modern technologies can have both positive and negative effects on production, society and the environment; for example, the use of robots can improve the efficiency of how products are put together on a factory assembly line. However, by replacing people, it can mean that fewer job opportunities are then available.

This section investigates the wider impact of new technologies and engineering industries. You will learn about some of the benefits and drawbacks of new and emerging technologies. You will also learn about how engineering industries can both help and cause problems for communities and the economy.

New and emerging technologies

New and emerging technologies are technologies that are either newly available or are in the process of being developed. There has been much advancement in the fields of robotics, nano-technology and communications in particular.

Impact on production

New and emerging technologies can improve the speed, efficiency and accuracy of product manufacture.

An example of this is the use of robotics to increase the amount of **automation** in production. Automation is the use of control systems to perform tasks, while reducing the amount of human involvement needed. Programmable robot arms can operate tools with a high degree of accuracy and precision. Unlike people, they do not need to take breaks, so they can operate continuously if required. In addition, there is a much smaller chance of human errors being made; this is only possible if mistakes occur when entering the robot's control program. This means there is less need to hire skilled workers, thereby reducing costs. However, new jobs may need to be created to program and maintain the robots.

Another example is the increasing use of rapid prototyping equipment and 3D printers. These allow manufacturers to create high-quality prototypes that can be tested, evaluated and improved relatively quickly. This allows errors in the designs to be spotted and corrected before too much time, effort and money is spent.

> **KEY WORDS**
>
> **New and emerging technology**: a technology that is either newly available or in the process of being developed.
>
> **Automation**: the use of control systems to perform tasks to reduce human involvement.

Figure 5.1.1 **A robot arm being used to aid product manufacture**

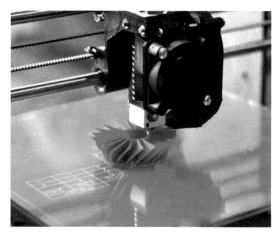

Figure 5.1.2 **A 3D printer being used to produce a prototype part**

Impact on society

New technologies can also have an impact on **society** so engineers have a moral responsibility to consider whether their use would help society or if it could be potentially damaging to it. This could be in terms of the impact on communities, culture, healthcare or on how large groups of people act or behave.

An example of this is the development of nano-technology. This is technology that exists on an 'molecular-scale' (extremely tiny scale), such as the size of atoms or molecules. Nano-robots, capable of entering the human bloodstream, have been used to destroy cancer cells in trials. Other potential benefits of this technology include improved fighting of infections, and even regrowing damaged organs. This could have a huge effect on how society experiences healthcare in the future. However, some people feel it is wrong to interact with the human body in this way.

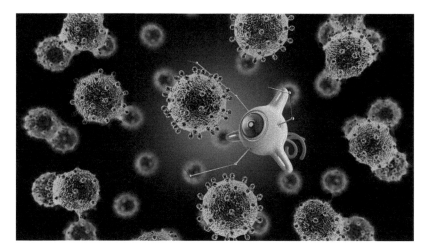

Figure 5.1.3 **Nano-technology can improve healthcare outcomes**

New and emerging technologies

Another example of a technology that could have a profound impact on society in the future is the 'internet of things' (IoT). This is the ability for everyday objects to be connected to the internet. It means that they can be controlled from any internet-enabled device, such as a tablet computer or a smartphone. In the future, you may not need to put the kettle or toaster on in the morning as your phone will have already done it for you!

Figure 5.1.4 **The internet of things**

STRETCH AND CHALLENGE

Produce a research report that explores example applications of nano-technology and new materials. What advantages may these modern technologies have on society? Are there any potential drawbacks?

You may wish to use the following as a starting point: https://science. howstuffworks.com/nanotechnology.htm

Impact on the environment

Engineered products require the sourcing of raw materials, which can be damaging to natural habitats. Some of these are **finite resources**, which means they cannot be replaced easily.

Sometimes, the effects on the **environment** are unintended or unplanned. For example, many new products are made using oil-based plastics. Once disposed of, much of this eventually ends up in Earth's oceans and the negative impacts of this type of waste have been huge and still not fully understood. For example, marine life is destroyed by eating the plastic, believing it to be food. This has resulted in the development of plastics made from natural resources and initiatives encouraging less use of plastic in general.

Pollution caused by the production, transportation and use of engineered products can also be damaging to the environment. **Carbon emissions** are directly linked to the greenhouse effect and global warming. The greenhouse effect is where heat is prevented from escaping from the Earth into space because of 'greenhouse gases' in the atmosphere (e.g. carbon dioxide), which, over time, are causing the Earth to become warmer.

New and emerging technologies can help to reduce carbon emissions by making greater use of 'cleaner' (less-polluting) energy sources. For example, replacing diesel engines with hybrid and/or electric vehicle engines significantly decreases the output of carbon dioxide into the atmosphere. It also has the effect of reducing the amount of oil that needs to be sourced and drilled for. In China, some buses use electric engines powered by a special type of capacitor, known as a super capacitor. These can hold much more electric charge than a regular capacitor, making them an ideal replacement for batteries. More information about capacitors can be found in Section 3.4.

Figure 5.1.5 **An electric car at a charging station**

KEY WORD

Fuel cells: power supplies that produce power from a chemical reaction between hydrogen and oxygen.

Fuel cells are another way of decreasing pollution and the amount of non-renewable energy used. These produce power from a chemical reaction between hydrogen and oxygen and the only by-product is water which does not harm the atmosphere. These can be used to provide power for buildings and transportation systems. They were originally developed to power spacecraft.

Further development of these technologies will become very important in the future as more countries, including the UK, move to permanently replace petrol- and diesel-powered forms of transport.

The use of smartphone technology can also help to reduce home energy usage. Specially designed apps allow people to control their lighting and heating systems from anywhere, ensuring that energy is not wasted when they are not at home.

ACTIVITY

Research and develop an engineered product that would have a positive impact on the environment. Your product must make use of at least one new or emerging technology.

Engineering industries

The term 'engineering industry' covers all aspects of the development and production of engineered products and systems. Engineers should be aware of the impacts of their work on society and the economy.

Impact on society

Engineering industries can impact on society in different ways, from improving opportunities for local communities to changing how people behave on a mass scale.

Figure 5.1.6 A mobile smartphone

An example of both of these is the mobile communications technology industry. The growth of this industry has revolutionised how people communicate with each other. In additional to voice calls, people can use text messages, video calls, emails and social media to stay in touch with their friends, family and colleagues instantly across the globe. They can conduct meetings with potential customers while hundreds of miles apart, reducing the cost of travel and opening up more business opportunities. Ease of communication has also allowed communities to be brought closer together. However, as people are making more use of text- or on screen-based communication, there is less need to talk in person and we must consider the negative impact this has on us socially.

Another social impact of engineering industries is on employment. According to EngineeringUK (www.engineeringuk.com), nearly 5.7 million people were working in engineering enterprises in the UK, in 2015. The more these businesses grow, the more opportunities become available for work. In addition to engineers and technicians, engineering companies also require people to work in finance, marketing and human resources departments. An engineering company based in an area can therefore be a very good source of work, providing for a wide range of roles and skills in the local community.

> **ACTIVITY**
>
> Identify a type of engineering industry that had an impact on society. Discuss the positives and negatives of this industry for society. You should consider both advantages and disadvantages in your discussion.

Impact on the economy

Engineering adds hundreds of billions of pounds in value to the UK economy every year, making it one of the largest sectors. There are hundreds of thousands of engineering enterprises in the UK and this number is growing. A successful engineering sector is therefore vital to the current and future prosperity (wealth) of the country. This is not just the case in the UK; for example, China's rapid economic growth in recent years has created many opportunities for engineering industries to thrive.

> **STRETCH AND CHALLENGE**
>
> Find an example of a country with a rapidly growing engineering sector. Research what is driving this and the outcomes for the country's economy.

Successfully putting engineering projects into use in society can also have benefits for the economy. A good example of this is improvements made to the transport infrastructure (the transport facilities serving an area or country). The London Crossrail project is estimated to put £42 billion into the UK economy through additional jobs, better connectivity and reduced journey times. The new high-speed railway, known as HS2, is predicted to bring even greater benefits over time (www.crossrail.co.uk/route/wider-economic-benefits, HS2: www.bbc.co.uk/

news/business-24040674). However, such projects can result in the loss of existing businesses due to the need for land and space. This can damage the local economies (economies of a town, city or other smaller area) of the communities that are affected.

Figure 5.1.7 **A section of Crossrail while under construction**

STRETCH AND CHALLENGE

Imagine that a high-speed railway is planned to go through your local town, connecting it with other major cities in the UK. Write a report analysing the potential economic and social benefits or drawbacks of this planned rail route. You should consider potential impacts at both a local and national level in your report.

KEY POINTS

- New and emerging technologies can improve production speed, efficiency and accuracy of production, but the increased use of automation and robotics may lead to job losses as machines perform tasks previously carried out by people.
- New and emerging technologies such as mobile technology make it easier for people to communicate with one another, but communicating via text or screen may reduce face-to-face communication and make us feel increasingly isolated.
- New and emerging technologies can have a positive impact on the environment, such as reducing reliance on fossil fuels and making use of 'cleaner' energy. However, the development of new products uses valuable raw materials that may have a negative impact on the environment when disposed of at the end of their useful life.
- Engineering industries can provide skilled job opportunities for people, but the development of more efficient automated systems may cause job losses for less skilled workers.
- Engineering industries contribute heavily to the success of the UK economy, such as through the completion of projects that improve infrastructure; but large infrastructure projects may lead to the loss of existing businesses and affecting the local economy.

Check your knowledge and understanding

1 Define the term 'new or emerging technology'.
2 Describe the positive effects that new and emerging technologies can have on the environment.
3 Explain how important the engineering sector is to the UK economy.

PRACTICE QUESTIONS : the impact of modern technologies

1 Which of these is a new or emerging technology?
   a Bearing
   b Chain and sprocket
   c Nano-robot
   d Spur-gear train

2 Which of the following is the most likely to occur as a result of increased production automation?
   a Decreased production efficiency
   b Decreased need for manual labour jobs
   c Decreased number of products made
   d Decreased production speed

3 Explain how one new or emerging technology can be used to improve production.

4 Give an example of a new or emerging technology that has had a negative impact on society. Explain this impact.

5 Discuss the statement 'Engineering industries have both a positive and a negative impact on the UK economy'.

6 Practical engineering skills

When manufacturing your product, you must apply the knowledge and understanding of engineering equipment, processes and mathematical understanding you have learnt so far. This should include safely using a range of tools and equipment; being able to select appropriate materials and components; and planning and executing the correct manufacturing processes.

This section will cover the practical engineering skills you will need to apply to produce your final working solution:

- producing a detailed design idea using appropriate engineering drawings that comply with industry standards and conventions
- producing and following a production plan that explains the stages of production and considers the quality control techniques used to safely produce the product
- using mathematical concepts and skills to predict performance of materials and working parts
- selecting and safely using appropriate materials, parts, components, tools and equipment to make a working product
- testing materials for suitability at the initial stages of manufacture and designing appropriate tests for a complete product to ensure it is fit for purpose and performance, suggesting improvements and modifications where appropriate.

This section includes the following topics:

6.1	Problem solving
6.2	Engineering drawings and schematics
6.3	CAD, CAM and CNC
6.4	Testing materials
6.5	Production plans
6.6	Predict performance using calculations and modelling
6.7	Select and use materials, parts, components, tools and equipment
6.8	Select and use appropriate processes
6.9	Apply quality control methods and techniques
6.10	Design tests to assess fitness for purpose and performance

This section is closely linked to the knowledge covered in Section 4, Testing and investigation.

Problem solving

What will I learn?

By the end of this chapter you should have developed knowledge and understanding of:

→ how to approach problem solving in a logical and systematic manner

→ how to analyse existing engineering solutions.

Problem solving is central to the work of all engineers. They design, manufacture, repair and maintain solutions to everyday living, which involves solving problems at all stages.

This chapter explains the main stages employed when solving engineering problems, and how solving problems can be done systematically.

It also looks at the importance of analysing existing engineering solutions and what can be learnt from doing this; for example, looking at how a mechanical system operates in order to improve its efficiency.

Make sure you have read and understood Section 3.1 Describing systems (pages 68–71) first.

___

Using a systematic approach to solve problems

As an engineer, you will need to identify and find solutions for a whole range of problems involving faults, failures, or ways to improve existing systems and methods. This should be done using a **logical and systematic approach**.

● The first stage of problem solving is to clearly understand and define (describe the nature of) the problem. If the problem is not fully understood, it will be almost impossible to solve it.

● If there are clients or other **stakeholders** involved, then their needs must be discussed. For example, what is the budget for the project? How must the finished product function?

● A short statement is sometimes written to clarify what the problem is, who the solution will be for and why it is important.

Example problem statement

Engineers working in factory environments sometimes have to move items of heavy equipment around the factory floor. This can be difficult or even impossible without the help of supporting equipment.

● The task is to design and manufacture a prototype scale model for a device that could lift different loads and move them to different places. The system should make use of mechanical, pneumatic and/or electronic systems as appropriate. It should also be sustainably powered.

● The next stage is to come up with alternative ways of solving the problem. System block diagrams (see page 68) can be drawn to outline ideas for potential solutions. These are quick to produce and do not go into too much technical detail. That makes them ideal for first ideas, and if they are ultimately discarded, then little time has been wasted.

● Once a number of possible first ideas have been considered, they are evaluated and narrowed down to the most appropriate. This is based on the evaluations of each possible solution and what will best meet the needs of the client, stakeholders, budget, etc.

- Schematics and flow charts may be created to show in more depth how the chosen solution would work. Flow charts show the order of operations in the system, whereas a schematic shows how all the component parts would connect together.
- The chosen solution is implemented (put to use) and its effectiveness is evaluated.

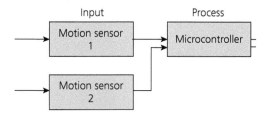

Figure 6.1.1 A block system diagram for a programmable security system

KEY WORDS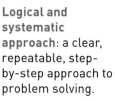

Logical and systematic approach: a clear, repeatable, step-by-step approach to problem solving.

Stakeholder: a person, group or organisation with an interest in the problem.

Functional characteristics: the features of a product or system that define how it works.

Analysing and investigating existing solutions

Engineers can learn a lot from analysing the work of others. Solutions to similar problems may already exist that can be developed further. As the saying goes; 'there is no point in reinventing the wheel'. As well as looking at what has previously worked well, engineers can also learn from past mistakes. This helps them to avoid making those same errors in their own solutions. For example, when it first opened, the London Millennium footbridge suffered from unexpected swaying, making it unsafe to walk on. It took nearly two years for the design to be fixed.

When analysing an existing system, engineers should think about:
- how it works
- how well it solved the engineering problem
- what could be improved or developed further.

This could also include the **functional characteristics** of individual components and sub-systems, safety issues and the impact on the environment. Drawing a block diagram of the system can help with this.

KEY POINTS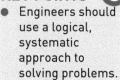
- Engineers should use a logical, systematic approach to solving problems.
- System block diagrams and flow charts are useful tools when solving engineering problems.
- When analysing existing solutions, engineers should consider what works well and what could be improved.

> **ACTIVITY**
>
> Conduct a detailed analysis of an existing engineering product or system that you have studied.
>
> This should include:
> - what sub-systems or component parts make up the product or system
> - how well the product or system functions
> - any safety considerations regarding the product
> - the impact it has or has had on the environment
> - any improvements that could be made to it.

Check your knowledge and understanding

1 Describe the main stages of solving an engineering problem.
2 Explain why engineers analyse existing solutions to problems.

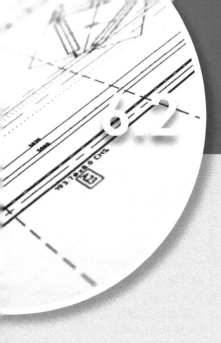

Engineering drawings and schematics

What will I learn?

By the end of this chapter you should have developed knowledge and understanding of:

→ how to produce and work to a series of mechanical, electrical or electronic engineering drawings or schematics using current conventions, including:
- orthographic (3rd angle)
- 3D representation (isometric)
- assembly
- sectional view.

An **engineering drawing**, also known as a working drawing, is a detailed drawing which gives instructions about how a product should be made. To produce an engineering drawing, you will need equipment such as:

- a drawing board
- T-square (for drawing horizontal lines)
- different kinds of drawing instruments such as:
  - different grade pencils (2H pencil to draw construction lines and a HB pencil to go around the outline of a drawing to make it stand out)
  - compass
  - set squares (60°–30° and 45°)

Computer aided design (CAD) software is now more commonly used for this application, as it is quicker and you can more easily create and modify engineering drawings compared to traditional methods.

There are different kinds of drawing methods used by engineers, all of which have to be drawn in accordance with British Standards (in this case PP 8888-1:2007 – Drawing practice: a guide for schools and colleges. See www.bsigroup.com for more information). This standard is a common method of producing drawings which everyone can follow (anybody can understand the drawing even if they do not speak English).

Schematic drawings

Schematic drawings are used by engineers to build electrical, hydraulic and pneumatic systems (refer back to Section 3, Systems). The most commonly used schematic drawing in engineering is the one used by electrical engineers to build electronic circuits. The schematic drawings are different to assembly drawings as they use standard symbols instead of drawing the parts. The drawing uses straight lines to connect these symbols together so they make a functional circuit. Producing the drawings this way means it is easy for an electrical engineer to interpret, follow and understand the intended product.

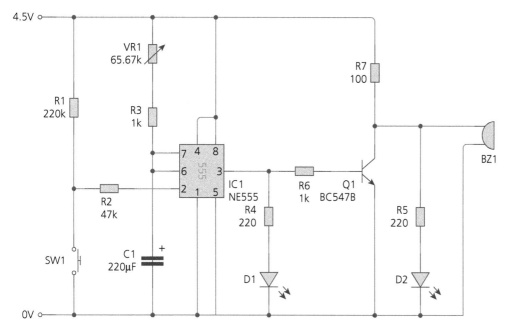

Figure 6.2.1 A schematic circuit diagram

The advantages of using schematic drawings are as follows:
● They are easy to understand as they use symbols and straight lines to connect individual parts.
● They use standard symbols with component values.
● The drawing can be followed by anyone from around the world as they do not need to be translated into another language.
● The drawing can be easily modified if additional components are required, or they can be added or changed later.
● A wiring diagram, which shows how the electrical wires are to be connected to electrical components, can easily be produced from a schematic drawing, and shows more details of the types of wires to be used (colour coded), the location of individual components on the circuit board, and how they are connected together to make a circuit.

Types of lines used for drawing

To make sure the drawing is easy to follow and understand, different lines and line thicknesses are used. The table and the image below show the most common ones found in engineering drawings.

——————	Thick line used for outlines of shapes and components to make them stand out
—————	Thin line used for construction, hatching and dimension lines
- - - - - - -	Dash line used to show hidden detail in a drawing
·—·—·—·—	Chain lines used for centre lines
A ⊥————⊥ A	Chain lines with thick ends and wide, filled-in arrow heads used to show position of sectioning

Table 6.2.1 Types of lines used for drawings

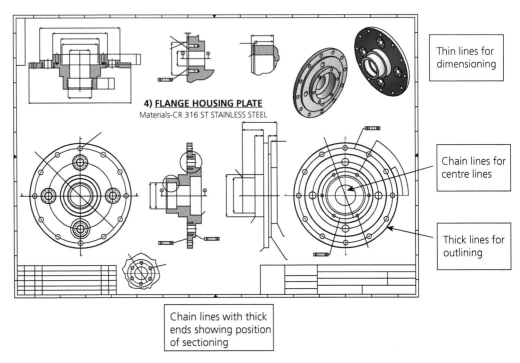

Figure 6.2.2 **How types of lines are used in engineering drawings**

Title block and parts list

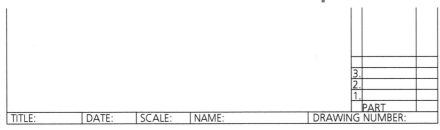

Figure 6.2.3 **Example of title block and parts list**

- the drawing number
- the scale of the drawing.

The **title block** is normally located at the bottom of the drawing with the parts list located on the bottom right hand side of the drawing.

The title block should include the following information:

- the title of the drawing
- the name of the person who did the drawing
- the date of the drawing

Dimensioning

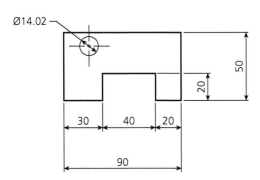

Figure 6.2.4 **Dimensioning is adding the sizes of the drawn part into the drawing**

The symbol used for **dimensioning** is shown in Figure 6.2.4.:

The following rules should be followed when dimensioning all types of engineering drawings:

- Do not dimension inside the object.
- The largest dimensions are put furthest away from the object.
- Dimension projection lines stop 2 mm away from the object.
- Dimensions are given in mm and numbers only.
- Dimension lines should be half the thickness of the object's outline.
- The dimension lines should be parallel to the object.
- Arrow heads should be filled in and be long and thin.
- Numbers should be placed above the middle of the dimension line.
- Dimension should be placed on the drawing so it can be read from the bottom or right-hand side.

Orthographic (3rd angle) drawing

Orthographic (3rd angle) drawing is the most commonly used method for drawing objects. The symbol used for the third angle projection is shown in Figure 6.2.5.

Orthographic (3rd angle) drawing gives all the information via three viewing points:
- plan
- front elevation
- side elevation.

These three viewing points are used to make the product – the figure below shows an orthographic projection drawing. In industry, two types of orthographic drawings are used for manufacture.

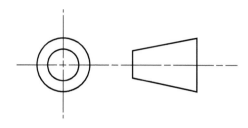

Figure 6.2.5 Symbol used for third angle projection

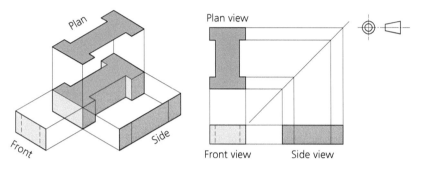 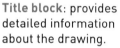

Figure 6.2.6 The different views of an orthographic projection are obtained from an object

General arrangement drawing

With a general arrangement drawing, the product is shown in its final assembled state. The overall dimensions are shown and each part is numbered.

Detailed drawings

Detailed drawings are a series of drawings of each component part. These show full dimensions, including working tolerances, and specified materials and finishes. They are attached to the general arrangement drawing.

Both sets of drawings are sent out together to manufacturers so they have full access to all the information required for manufacturing.

Produce the orthographic (3rd angle) projection drawing

To produce an orthographic (3rd angle) projection drawing you must:

- Step 1 – Draw the title block and add drawing information. Set up the page making sure all three viewing points will fit onto the page. The plan view is a very good starting point so this needs to be drawn towards the top left-hand side of the paper. The outline of the view can be outlined with a HB pencil to make it stand out.

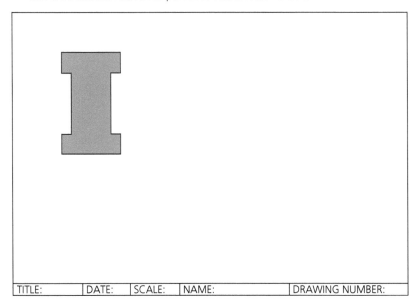

Figure 6.2.7a

- Step 2 – Project the construction line down from the plan view. A 2H pencil can be used for drawing these. The construction lines are going to be used to draw the front-view detail in. Remember to project the hidden detail into the front view, which is shown with a dotted line. The outline of the view can be outlined with a HB pencil.

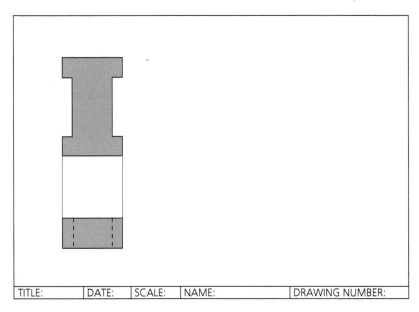

Figure 6.2.7b

● Step 3 – From the top corner of the front view, project a 45° line using a 2H pencil, as this line is going to be used as a construction line.

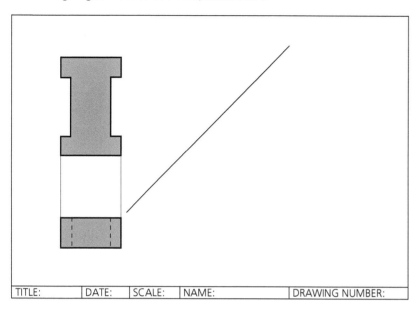

TITLE:	DATE:	SCALE:	NAME:	DRAWING NUMBER:

Figure 6.2.7c

● Step 4 – Project construction lines from the plan view to the 45° construction line that you have just drawn. These are going to be used to construct the side-view projection.

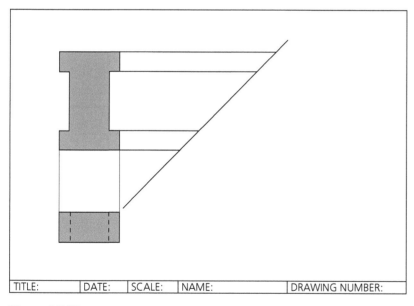

TITLE:	DATE:	SCALE:	NAME:	DRAWING NUMBER:

Figure 6.2.7d

Orthographic (3rd angle) drawing

- Step 5 – Project lines down from the 45° line and across from the front view; you have now produced the side-view projection.

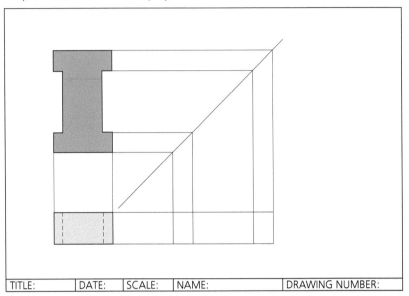

Figure 6.2.7e

- Step 6 – Using a HB pencil, outline the side-view projection to make it stand out. You have now completed the third angle projection drawing. All there is left to do is add the dimensions to the drawing and the third angle projection symbol.

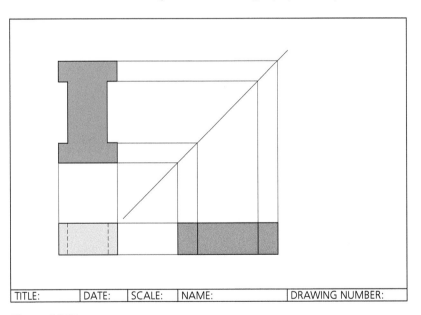

Figure 6.2.7f

- Step 7 – Add the dimensions correctly to the drawing. Remember the rules about dimensioning a drawing correctly.

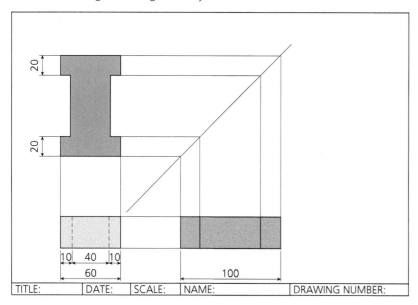

Figure 6.2.7g

- Step 8 – Complete the drawing by adding the third angle projection symbol.

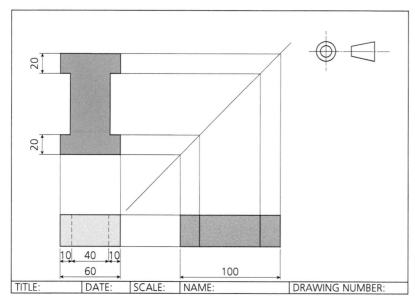

Figure 6.2.7h

Orthographic (3rd angle) drawing

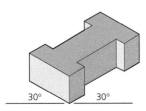
3D representation (isometric) projection drawing

A **3D representation (isometric) drawing** gives views of three sides of an object. It is a very accurate way of drawing 3D shapes, especially if using gridded paper as shown below. The gridded paper can be used to draw directly onto or can be placed behind your drawing paper so it shows through, giving a clear reference where to draw.

Drawing isometric projection

To produce a 3D representation (isometric) drawing you must:
- draw all lines to scale (all measurements are actual size)
- ensure lines that are vertical on the object are also kept vertical in the drawing
- ensure all lines on the drawing (except the vertical lines) are drawn at 30°.

Figure 6.2.8 An object drawn in isometric projection

Assembly drawings

Assembly drawings (also called general arrangement drawings) are used when many different parts of a product come together. They show exactly how the product should be fitted together and are used mainly by toolmakers and machine shop fitters to help with product assembly.

The figure below shows different parts of an assembly being produced by CAD.

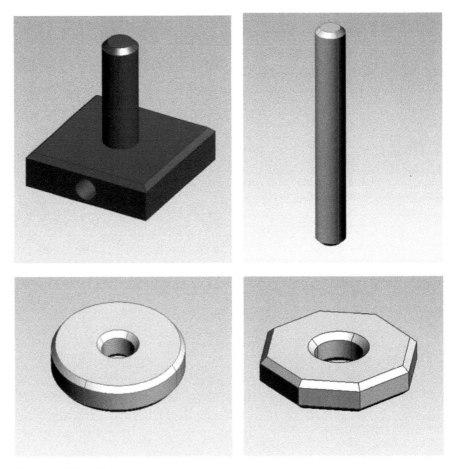

Figure 6.2.9 **CAD designs of parts**

Figure 6.2.10 shows how all the parts have come together as an assembly.

Once the designer is happy with the assembled design, CAD software can be used to transform it into an orthographic assembly drawing as shown below.

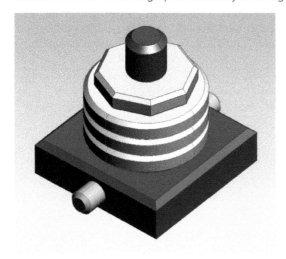

Figure 6.2.10 CAD assembly of the parts

Information that can be used in an assembly drawing

The following information is often used in an assembly drawing:
- information on how to assemble the product
- assembled outside/overall dimensions
- surface-finish requirements
- a full list of parts which relate to labelled parts on the drawing.

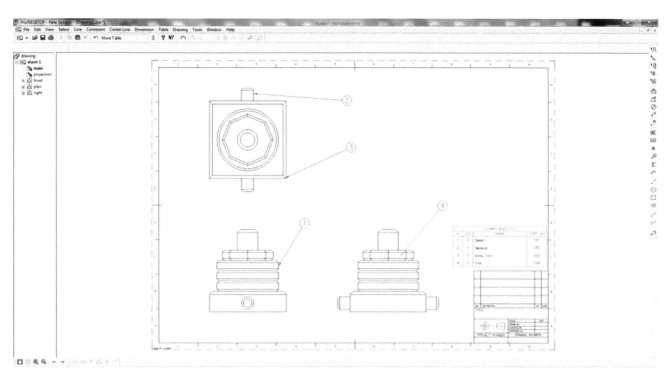

Figure 6.2.11 CAD assembly drawing of the components in 3rd angle orthographic projection

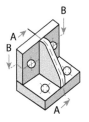
Section view

Section-view drawings are views of the inside of an object as if it has been sliced open through two or more points or 'planes' of the object. This process is called sectioning and shows detail of what a product or component would look like inside. Different parts are shown cross-hatched (equally spaced lines drawn at 45°). Each part is cross-hatched in different directions, making assembly easy to understand. Screws, nuts and bolts and webs (additional material added to a component to help to strengthen its structure) are not normally sectioned. No hidden detail (i.e. features of an object which are hidden inside a component and cannot be seen from a particular view) is included in a section-view drawing.

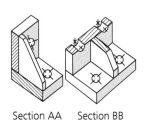

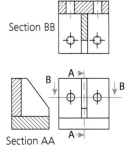

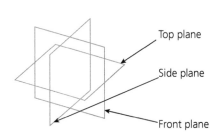

Figure 6.2.12 Sectioning of a component

Figure 6.2.13 The three planes of a drawing

STRETCH AND CHALLENGE

Produce an isometric drawing of the 13 amp plug.

KEY POINTS

- Thick lines are used for outlining the outside of a part to make it stand out; thin lines are used for construction lines, dimension lines, and for hatching in a sectional view.
- A dash line is used to show hidden detail in a drawing.
- Chain lines are used as centre lines. A chain line with two thick ends and an arrow pointing to each one with a letter is used to show positioning of a sectional view.
- Orthographic drawings have three viewing points.
- Isometric drawings are drawn with all lines to scale (all measurements are actual size) and are drawn at 30°, with the exception of vertical lines on the object which are kept vertical in the drawing.
- Assembly drawings are used to show how different parts of a product should be fitted together.
- Section views provide a view of the inside of an object as if it has been sliced open.

Check your knowledge and understanding

1 Identify the grade of pencil that should be used to outline your drawing work.
2 Draw the symbol for third angle projection.
3 Describe what information is required in the title block.
4 Describe what information could be used in an assembly drawing.
5 Explain why schematic drawings are easy to follow and understand.

6.3

CAD, CAM and CNC

What will I learn?

By the end of this chapter you should have developed knowledge and understanding of:

→ how computer aided design (CAD) is used to produce 2D and 3D drawings to assist in the creation of a solution

→ how 2D computer aided manufacture (CAM) is used to control a laser cutter, vinyl cutting, PCB manufacture and CNC turning.

→ how 3D computer aided manufacture (CAM) is used in rapid prototyping and milling/routing.

Using CAD to create 2D and 3D drawings

Computer aided design (CAD) is computer software used by a designer to design, mould and produce full working drawings such as orthographic drawings. CAD programs can also be used to produce schematic drawings for electrical, electronic and pneumatic systems. The software can select components from a library and add them into the drawing, linking them with connectors. The software can 'dry run' (simulate the system) to see if it will function.

Using CAD to create 2D and 3D drawings allows very accurate designs to be produced. Modifications can be made to a drawing very easily and quickly without having to start from the beginning.

Schematic drawings

The operating parameters can be entered into the CAD package. The software will then work out and select suitable components that could be used in the drawing. The designer can select these components to produce electrical, electronic and pneumatic schematic drawings. An example of this is shown in Figure 6.3.1 in which a 2D circuit diagram has been produced. From here the software has generated a printed circuit board (PCB) which can be used to produce the circuit.

Figure 6.3.1a **Schematic electrical CAD drawing**

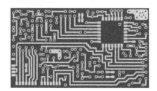

Figure 6.3.1b **PCB designed by CAD**

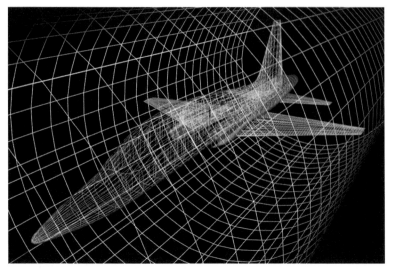

Figure 6.3.2 **An aircraft being tested in a virtual wind tunnel**

Virtual reality simulation

Systems/products can be modelled by the CAD software to produce a realistic 3D model of a product. Virtual reality simulation can be used to show the systems and products in operation. This will save a lot of money as prototype and testing rigs will not have to be produced. Aircraft engineers use virtual reality simulation to test new designs. The testing of an aircraft in a wind tunnel simulation is shown in Figure 6.3.2.

Orthographic drawings

These can be produced by the CAD software, which can be sent directly to external manufacturers or different manufacturing departments within the factory.

Full rendered assembly drawings

These can be produced to show how products will look. The image below shows a model of a rendered drawing.

Figure 6.3.3 **A 3D rendered drawing of a car**

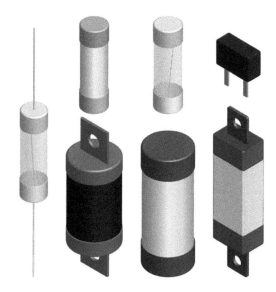

Figure 6.3.4 **Isometric view of electronic components**

Isometric and sectional views

Isometric and sectional views of a product can easily be produced on screen with the CAD software. Figure 6.3.4 shows an isometric view of electronic components.

Using CAD with CAM

CAD drawings can be sent directly to computer aided manufacturing (CAM).

Computer aided manufacture (CAM)

Computer aided manufacture (CAM) is when computers are used to control and operate machinery. CAM is used across all sections of engineering. All CAM machines use **computer numerical control (CNC)**.

Computer numerical control (CNC)

CNC is controlled by a number system which is known as machine code. The machine code controls the speed and feed rate (the speed at which the work is presented to the cutter or the tool is moved by the machine) of the cutter in the *x, y, z* directions for horizontal machines and in the *x, y* directions for lathes. The machine code (simple binary language used by computers) is stored in a machine; this is known as a program. Usually CAD software can be used to produce the program for the CAM which is used to machine the components.

Figure 6.3.5 **A CNC milling machine which uses computer numerical control**

Computer aided manufacture (CAM) 2D

With CAM 2D, the cutting tools can only move in two axes (*x, z*). Figure 6.3.6 shows an example of a two-axis CAM machine. The parts to be cut out are designed using CAD software. When the design is completed, the settings of the CAM machine can be adjusted on the CAD software via the computer which translates the design into machine code. The code is then sent to the CAM machine, which stores it as a program. The program can be run by the CAM machine which cuts or machines the design.

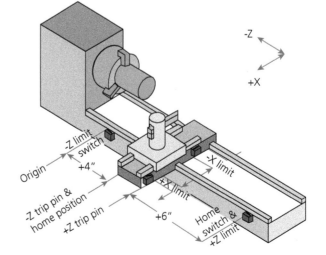

Figure 6.3.6 **A two-axis CAM lathe**

Figure 6.3.7 **A CAM 2D laser cutter**

Examples of CAM 2D machinery include the following:

Laser cutter

A laser cutter works by directing a high output laser through a range of optics. The focused beam is used along with the CNC control to engrave and cut material.

Vinyl printer/cutter

As the image below shows, the vinyl printer/cutter prints the image directly onto thin self-adhesive backed vinyl. The image is then cut out using a very sharp cutting blade which is held in a CNC-controlled tool holder. The vinyl cutter can be used without the printer head to cut out different shapes in different coloured vinyl.

Figure 6.3.8 **CAM 2D vinyl printer/cutter**

Figure 6.3.9 **Close-up view of the tool holder and blade**

CAM printed circuit board (PCB) manufacture

Copper patterning is used to print a protective layer on top of the proposed track-layout design on the PCB board (the PCB board is made up of thin copper layers). Etching (the removing of an unprotected metal by acid) is then used to remove the unprotected copper layers, which leaves the initial protected copper track. CNC drilling can then be used to drill holes in the track for the electronic components.

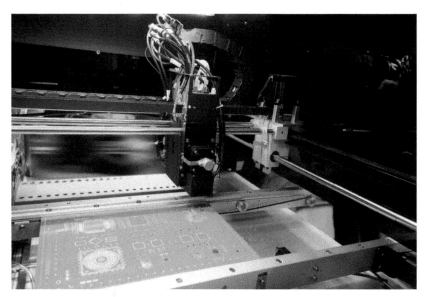

Figure 6.3.10 **PCB manufacture**

Computer numerical control (CNC) lathe

The material is held in a vice which can be operated manually or automatically with the aid of a pneumatic chuck. The machine program is then run which operates the speed of the chuck rotation and the speed and feeds of the tool post in the *x, z* direction. The capstan tool post could contain up to ten different tools, which can be selected and used via the program. This is a very quick and extremely accurate way of producing cylindrical work.

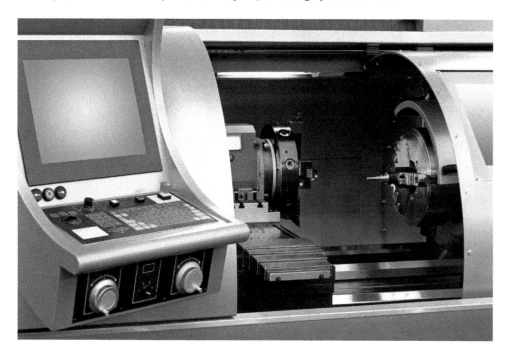

Figure 6.3.11 **CAM turning machine**

Computer aided manufacture (CAM) 3D

CAM 3D uses cutting and forming tooling which moves in 3 axes (*x, y, z*). As a result, 3D shapes can be cut and formed.

The CAM 3D uses different software to CAM 2D as it needs programming that can produce 3D images. The images are sent from the software to the CAM 3D as machine code which, like the CAM 2D, is stored as a program in the CAM 3D machine. The machine then runs the program and produces the required component.

Examples of CAM 3D machinery include the following:

3D printer

A 3D printer uses additive manufacturing to produce a three-dimensional component. The additive 3D printing technique used on 3D printers is called fused filament fabrication. 'Additive' refers to the molten plastic added in small amounts. A thin roll of thermoplastic wire is fed

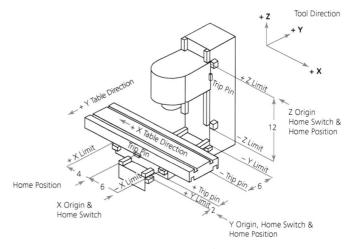

Figure 6.3.12 **Diagram showing a CAM 3D (3-axis machine)**

through a moving heating element that melts and **extrudes** (forces out) the plastic wire. It then deposits the molten plastic to form a layer of hardened plastic (it solidifies as it cools down). As each layer is completed, the bed of the machine moves down a layer (*z* axis) so the heating element can pass over again, depositing more material (the heating element moves on the *x–y* plane). The process carries on until the shape of the component has been produced.

Computer aided manufacture (CAM)

KEY WORD 🔑

Extrudes: the material is heated up and pushed through a die where it is deposited until it cools down.

Figure 6.3.13 **3D printer using additive manufacture**

CAM 3D milling and routing

CAM 3D milling and routing can perform multi-axis machining. This is when the CNC control can move the cutter in four or more ways, resulting in their being able to machine very complex shapes. They also have dynamic control of both the tooling and the work piece (the tooling and work piece can both move). The machines can be outfitted with a number of tool heads to machine different materials and profiles. The CNC program changes the tool heads automatically when required, so the whole manufacturing process is completely automated.

Figure 6.3.14 **A CAM 3D milling machine in operation**

Check your knowledge and understanding

1 Explain how CAM machines are controlled when manufacturing a product.
2 A complex 3D shape needs to be produced by a manufacturer. Suggest a piece of equipment to produce:
   i   a 3D prototype
   ii the product made from low-carbon mild steel.
3 Explain why virtual reality simulation is frequently used by designers.
4 Discuss the advantages of using CAD to produce a schematic drawing for a pneumatic system.

6.4 Testing materials

What will I learn?

By the end of this chapter you should have developed knowledge and understanding of:

→ methods of testing and evaluating materials and structural behaviour under load
→ types of non-destructive and destructive testing
→ testing tensile and compressive strength of different materials.

A materials test is used to check the working properties of a material before it goes into service or it can be used to identify an unknown material by its configuration (how its structure has been arranged and put together).

Methods for testing materials

Material testing can be put into two categories:

- non-destructive testing – in which the component or material can still be used after testing
- destructive testing – in which the component or material cannot be used after testing.

Non-destructive testing used for finding faults and defects in materials

Visual testing is a form of non-destructive testing. It involves using the human eye to visually inspect and test a product. As a result, it is the most commonly used test.

Visual testing is a very quick and effective way of checking components, including welds, forgings, machined components, castings, etc. It is done by the human eye so expensive test equipment is not needed to carry out the procedure. The test is useful as a quality control check on a production line. Usually, a pre-made test sample is used to compare, making it easier to identify any differences. Other forms of non-destructive testing include ultrasonic testing and radiography

Destructive testing to investigate the different properties of a material

The following are types of destructive testing to investigate the properties of materials:

- Tensile test – to find the strength of a material with a pulling force.
- Compressive strength test – to find the strength of a material with a pushing force.
- Hardness test – to find how hard a material is.

The form of destructive test used can depend on the type of material.

Testing materials can help engineers to find out the working properties of a material so it can be selected for a certain application. It can also help with deciding the best tooling and machining methods for the material.

Tensile test

The purpose of the test is to find the strength of the material under tension. The **tensile strength** of a material is the maximum pulling/stretching force it can withstand before failure. (See Chapter 4.2 for more information on tensile testing).

The test is carried out by making a specimen test piece out of the chosen material (as shown in the image below).

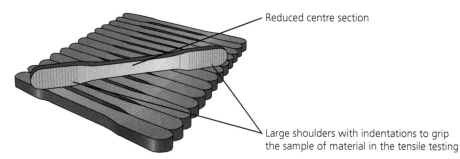

Reduced centre section

Large shoulders with indentations to grip the sample of material in the tensile testing

Figure 6.4.1 A set of sample test pieces

- The specimen is produced with two large shoulders which can be gripped by the tensile testing machine and a reduced centre section which is used for the testing.
- A force is then applied to the test piece by hand via a turn handle.
- The load being applied to the test piece is displayed via a pointer on a scale, with the force being increased as the handle is turned.
- During initial stages of the test, the material exhibits elastic proprieties (the material can be returned to its original shape if the load is removed).
- As more force is applied, the material reaches a stage where it stretches without any resistance. This point is called the yield point.
- The material recovers slightly but eventually becomes a lot thinner as more force is applied, until fracture occurs. This point is the tensile strength of the material.

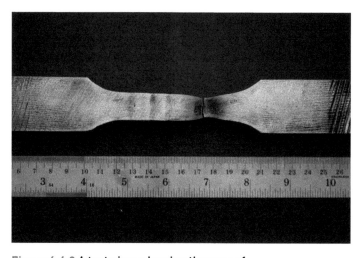

Figure 6.4.3 A test piece showing the area of fracture after testing has taken place

Figure 6.4.2 A tensile testing machine

Compressive strength test

Compressive strength is the resistance of a material to fail under a pushing/compressive force. The test to measure compressive strength is similar to a tensile test. However, instead of the test material being in a pulling force until breaking, it is placed in **compression** (being pushed together) instead. The test involves the test material resisting compression instead of being pulled apart as in the tensile test.

Metals tend to have similar compressive and tensile strengths. However, cast iron has a higher compressive strength than a tensile strength. Concrete has a much higher compressive strength than a tensile strength, which makes it ideal for applications under compression such as foundations for buildings.

Hardness test (Rockwell test)

The Rockwell hardness test works by measuring how deep an indentation is made by the testing machine as it forces an indenter (a small round bar which tapers to a ball end-point) under a load into the material being tested.

Figure 6.4.4 **A tungsten carbide indenter**

The indentation made by the indenter is measured against a scale (hardness values of different materials) which identifies how hard the material is.

KEY WORDS

Visual testing: inspecting a product with the human eye.

Tensile strength test: tests the maximum load a material can withstand without breaking.

Compression test: applying pressing forces to either side of an object.

Hardness (Rockwell) test: determines the hardness of a material

---

KEY POINTS
- Destructive testing involves applying a destructive test to a sample. The sample will be unable to be used after testing.
- Non-destructive testing involves applying a test to a sample. The sample can be used after testing.
- Tensile strength is the maximum pulling/stretching force a material can withstand before failure.
- Compressive strength testing is the resistance of a material to fail under a pushing/compressive force.
- Visual testing is a non-destructive test to quickly and effectively check components.

Check your knowledge and understanding

1 Explain why visual tests are carried out on products.
2 State the difference between destructive and non-destructive testing.
3 Explain how a tensile test is carried out on a sample of material.
4 Identify a test for finding the hardness of a material.

Production plans

What will I learn?

By the end of this chapter you should have developed knowledge and understanding of:

→ how to produce and follow production plans, taking into account materials, processes, time and safety
→ the importance of production control.

Production planning plays a very important part in manufacture. This is when the production process is planned in the most systematic way, to make sure the entire process is carried out in the most cost-effective way. This advanced planning ensures better quality goods at reasonable prices as production is meticulously planned.

Production planning in engineering

Before starting the manufacturing of a product, it is important to plan ahead so the production can run as smoothly and cost effectively as possible within the planned set time frame. In industry, a **production plan** is created to plan ahead.

Plan production schedules according to required quantities

The market and sales team will have forecast expected sales or they may need to release a new product to compete with competitors. They feed back to production planning which, in turn, schedules production runs of the new product. Smaller factories may have limited use of machinery resulting in them having to use the same machine to produce different articles. An example of this would be a 200-ton press. It may be needed to produce five different components in a one-week rotation. A production schedule would need to be used to plan an effective way of managing this issue to avoid as much downtime as possible.

Plan for different departments to work together

It is very important that different departments work together relating to production needs. In the case of a new product, the design department will need to work closely with the production engineering department, which may need to buy in new machinery and tooling to produce new parts. Production engineering will need to work closely with the tool room manufacture department which will need to produce new tooling, jigs and fixtures for the new production line.

Plan for different types of production methods

The quantity of the product to be manufactured will decide the scale of the production method required; from single and batch production to mass, or continuous production methods for very large quantities.

Plan the number of workers, materials and equipment required

So that production is cost effective and products are made to a tight deadline, it is important that the correct number of workers are employed to carry out manufacture; the correct amount of material is ordered, with discounts for large purchases; and the most cost-effective equipment is purchased or leased.

Plan for available materials, parts and components

Many manufacturing sectors use just-in-time delivery for materials, parts and components, as they do not need to heat and light large warehouses to store them in. However, it is important that everything arrives on time as missing or late delivery of materials, parts and components will cause production to stop, which is a very costly mistake for the manufacturer.

Plan for effective assembly line

All services which keep the main production line running effectively, without interruption, will need to be regularly checked and maintained to avoid unnecessary breaks in production runs.

Plan for backup materials, components and machinery

Unpredicted issues occur in manufacture from natural disasters such as flooding or snow where roads may be blocked so materials, components, etc. cannot reach the manufacturer. Machinery can also break. Emergency backups should be available on standby in case replacement parts, to fix existing machinery, cannot be sourced quickly. Whatever the problem, there should always be a backup plan in case of a production emergency.

Plan to produce cost-effective, high quality products

If production planning is carried out correctly, the production line should run effectively, producing high quality products at the least possible total cost. The more produced to budget, the cheaper the final cost of manufacture. These savings can then be passed onto the consumer who, because of the price, may buy more of the product boosting sales.

Plan to ensure products are produced on time

Ensure products are produced on time to avoid consumer disappointment.

Plan to manufacture in-house

Where possible, plan to manufacture everything involved in-house to utilise machinery and labour.

Production control

The **production control** process makes sure that all production planning targets are met by running the production control system of the whole factory effectively.

The advantages of effective production control are as follows:
- It ensures production runs smoothly and effectively.
- All production targets are met.
- The least amount of waste is produced by the factory when producing the product.
- High standard, quality products are produced.
- The item is manufactured for the correct price.

- Orders from the sales department are fulfilled on time.
- Production scheduling is carried out effectively.

Only with *both* effective production planning and production control in place will the customer receive a quality item.

Production plan for the manufacture of 20 location pins

The diagram shows the image of the location pin to be produced.

The school has allocated a time of 3 hours. They have allowed three metal turning lathes to be used for the manufacture which, in turn, are operated by three skilled pupil operators and one quality control checker. The production plan shows each operation, including the machinery and personnel required to carry out each operation. Time to carry out total manufacture = 3 hours.

Total number of personnel to carry out manufacture – 4.

3 x lathe operators, 1 x quality control checker.

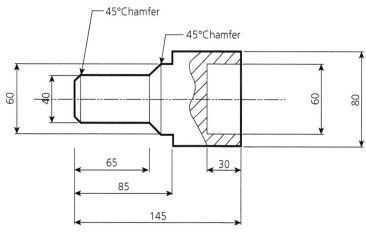

Figure 6.5.1 The location pin to be produced

Operation	Detail of operation to be carried out	Lathe 1 Operator 1	Lathe 2 Operator 2	Lathe 3 Operator 3	Quality Control Checker
1	Set up 60mm stock bar and face off	3 minutes			
2	Turn down to diameter 60mm	2 minutes			
3	Turn down to diameter 40mm	2 minutes			
4	Part off to length of 145mm	2 minutes			
5	Check part				1 minute
6	Set up and turn 2 x 45 degree chamfers		2 minutes		
7	Set up component drill and bore 60mm hole to required depth of 30mm			5 minutes	
8	Check part				1 minute

Table 6.5.1 Example production plan that could be used in schools to produce a batch of 20 location pins in the most cost-effective way

The production plan shows:

If a single person had been used to produce all pins it would have taken them

$$3 + 2 + 2 + 2 + 1 + 2 + 5 + 1 = 18 \text{ minutes per location pin.}$$
$$20 \text{ pins x } 18 \text{ minutes} = 6 \text{ hours to produce}$$

With 3 operators and 1 quality control checker the total production time can be reduced by 3 hours saving production time and improved quality control.

KEY WORDS

Production plan: a detailed plan related to the manufacturing process of a product, which helps ensure that the highest quality goods are produced safely, using materials and processes in a set time.

Production control: the measures taken to ensure production runs smoothly by following the production plan.

KEY POINTS

- Production planning is done beforehand to ensure the manufacturing of a product runs as smoothly and is as cost effective as possible within the planned set time frame and budget.
- The correct production method should be used for the quantity of product to be manufactured; for example, a continuous run is no good for small quantities of a product.
- Production control ensures planning targets are adopted and the production control system runs smoothly and effectively.

Check your knowledge and understanding

1 Explain why production planning is important.
2 Describe how departments work together relating to product needs.
3 Give two reasons why materials are bought in bulk.
4 Explain why it is important a manufacturer produces high quality products.
5 Explain how the production plan ensures quality products are produced.

6.6

Predict performance using calculations and modelling

What will I learn?

By the end of this chapter you should have developed knowledge and understanding of:

→ how to predict a product's performance using calculations
→ how to predict a product's performance using modelling such as iconic, analogue and symbolic modelling.

As introduced in Section 4.1 Modelling and calculating (see pages 106–123), engineering processes require you to apply your mathematical skills. This section will help you to further understand some important calculations used in engineering to assess processes such as stress, strain, density and mechanical advantage, and provide you with some examples to practically apply your mathematical skills.

You should also refer to the Engineering equations and Engineering symbols sections at the back of this textbook.

Units

Units refers to the measurement of a property of a material.

mm/m =	length
kg/tonnes =	weight
newtons =	force
volts =	voltage
ohms =	electrical resistance
m^2 =	area
cm^3 =	volume
litres	liquid

Table 6.6.1 **Units of measurement**

ACTIVITY

Fill the gaps in the following:

? mm = 1 cm   ? cm = 1 m
? g = 1 kg   ? Kg = 1 tonne
1 cm = 0.? m   1 mm = 0. ? cm

Degrees of accuracy

In engineering, it is very important to be as accurate and precise as possible when working out calculations. The final answer can be improved by using significant figures to improve its accuracy and precision.

Decimal places (d.p.)

Answers should be given to a set number of decimal places (d.p.); for example, 3.824 is 3.82 (to 2 d.p.).

Significant figures (sig.fig.)

Significant figures refers to answers given with only a set amount of figures; 0s are inserted to show the position of the decimal point.

MATHEMATICAL UNDERSTANDING

Significant figures example

1. Give 1364.2 to 2 and 3 significant figures.

2. Give 12.37 to 2 and 3 signification figures.

Solution

1. To 2 sig. fig. – answer is 1400

   To 3 sig. fig. – answer is 1360

2. To 2 sig. fig. – answer is 12.0

   To 3 sig. fig. – answer is 12.4

Conversion

Before completing calculations, you must take note of the units of measure. You may need to convert or change the form of one or more sets of measurements first.

MATHEMATICAL UNDERSTANDING

Conversion example

A sheet of low-carbon mild steel measures seventy-five centimetres by ninety-five centimetres.

Calculate its area in m^2.

Solution

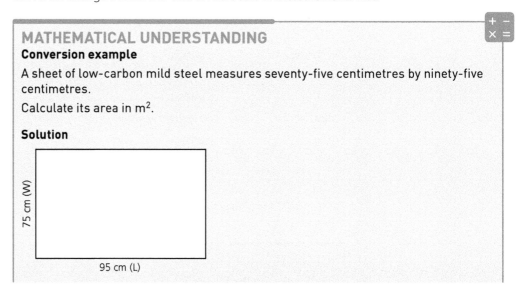

There are two methods. Both give the same answer.

Method 1	Method 2
Work in metres throughout.	Work in centimetres then convert to m^2 at the end.
W = 75 cm = 0.75 m	W = 75 cm
L = 95 cm = 0.95 m	L = 95 cm
A = L x W	A = L x W
= 0.95 x 0.75	= 75 x 95
= 0.7125 m^2	= 7125 cm^2
= 0.71 m^2 (to 2 d.p.)	= 0.7125 m^2
	= 0.71 m^2 (to 2 d.p.).

As the question does not ask for a set degree of accuracy, choose your own.

To convert cm^2 into m^2

$$1 \text{ metre} = 100cm$$

$$1 \text{ } m^2 = 100cm \times 100cm$$

$$1 \text{ } m^2 = 10\,000cm^2$$

So to convert cm^2 into m^2 we need to divide by 10 000.

This is the same as moving the decimal point 4 places to the left.

ACTIVITY

Calculate the following conversions:

1 A metal bar has a rectangular cross section which measures 90 cm by 75 cm, and is 1.5 metres long. Calculate its volume correct to the nearest m^3.

2 A circle has a radius of 7 cm. Calculate its area correct to the nearest whole number.

3 A cylindrical metal bar is 3 metres long and has a radius of 14 cm. Calculate its volume in m^3. Give your answer correct to 2 significant figures.

Standard form

Standard form helps to write big or small long numbers so that calculation/working out can be displayed more easily.

You will have used this in science and maths. Here is a reminder.

MATHEMATICAL UNDERSTANDING

Standard form example

In standard form numbers are written as powers of ten; for example, $530 = 5.3 \times 10^2$

This number is always less than 10
(even if the original number is less than 1).

ACTIVITY

Write in standard form:

1 3862
2 5601
3 4169
4 50

Smaller numbers are written in the same way but with a negative power, for example, 0.083 is written as 8.3×10^{-2}

You will notice that we find the power by counting the number of decimal places the decimal point moves.

Try these:

5 0.0072
6 0.00076
7 0.84

You are often asked to give answers in standard form. Here are some examples which use some of the formula you need to know.

Stress

We calculate stress to find a material's safe working limit. When this is exceeded the material will fail. Calculating stress in a material will help engineers decide if the material is suitable for an engineering application.

MATHEMATICAL UNDERSTANDING

Stress example

A rod has a circular cross section of diameter 35 m. Calculate the stress in the tie rod when it is subjected to a load of 30 000 N. Give your answer in standard form.

Solution

Use formula: $\text{Stress} = \dfrac{\text{Force}}{\text{Cross-sectional area}}$

$$F = 30\,000 \text{ N}$$
$$\text{Diameter} = 35 \text{ m}$$
$$\text{Radius} = 17.5$$
$$\text{Area of cross section} = \pi r^2 \qquad = \pi \times (17.5)^2$$
$$\text{Stress} = \frac{30\,000}{\pi \times (17.5)^2} = 3.1 \times 10$$

Strain

It is often necessary to calculate the deformation (how much it expands) of a solid material due to the amount of load applied to it. Most materials are elastic and will return to their original length when the load is removed.

MATHEMATICAL UNDERSTANDING

Strain example

Calculate the strain when a bar of original length 5 m stretches by 0.25 mm. Give your answers in standard form.

Solution

$$\text{Strain} = \frac{\text{Change in length}}{\text{Original length}} = \frac{0.25 \text{ mm}}{5 \text{ m}} = \frac{0.25 \text{ mm}}{5000 \text{ mm}}$$

$$= 0.00005$$
$$\text{Strain} = 5.0 \times 10^{-5}$$

Density

Density measures how compact an object is. When materials such as iron change from a solid state to a liquid state as they are melted by heat, their density does not change very much because the particles of the material stay the same distance apart in different states.

We measure density in $kg\ m^{-3}$, kg/m^3 or g/cm^3.

Formula: $Density = \dfrac{Mass}{Volume}$

MATHEMATICAL UNDERSTANDING

Density example

Rocks are used to prevent coastal erosion and are made from a material of density $3500\ kg/m^3$, and will need to have a mass of 2 tonnes each. What volume will one rock cover?

Solution

$$Density = \frac{Mass}{Volume}$$

$$Density = 3500 = \frac{2000}{Volume}$$

$$Mass = 2\ tonnes = 2000\ kg$$

$$Volume \times 3500 = 2000$$

$$Volume = \frac{2000}{3500}$$

$$= 0.571\ kg/m^3$$

$$Volume = 5.7 \times 10^{-1}$$

Young's Modulus

Young's Modulus is the measure of the stiffness and elasticity of a material (see Section 4.1 Modelling and calculating).

In engineering, one of the most important tests is knowing when an object or material will bend or break. Young's Modulus measures how easily a material stretches.

We calculate Young's Modulus using the following formula:

$Young's\ Modulus = \dfrac{Stress}{Strain}$

The higher the value of Young's Modulus, the lower are the chances of breakage (the stronger the material), i.e. the material is safer to use for the task. There have already been examples of calculating stress and strain.

MATHEMATICAL UNDERSTANDING

Young's Modulus example

Calculate Young's Modulus if

$$Stress = 2.27$$

and $strain = 0.2$

Solution

$$Young's\ Modulus = \frac{2.27}{0.2} = 11.35$$

$$= 11.4\ (to\ 1\ d.p.)$$ The number has been made simpler but has still kept its accuracy.

Ohm's Law

Ohm's Law is the relationship between voltage, current and resistance in an electrical circuit.

Resistance is measured in ohms (Ω).

The greater the number of ohms, the greater the resistance.

$$\text{Current} = \frac{\text{Voltage}}{\text{Resistance}}$$

We could also write this as voltage (volt, V) = current (amperes, A) $\times$ resistance (ohms, Ω).

MATHEMATICAL UNDERSTANDING
Ohm's Law example

A bicycle has a battery-operated light. The bulb in the rear light has a resistance of 5 Ω and takes a current of 0.5 A. At what voltage does it work?

Solution

Current = 0.5

Resistance = 5

Voltage = Current $\times$ resistance

$= 0.5 \times 5$

Voltage = 2.5 volts

In the above example, there is also a front lamp of resistance 2 Ω which also needs a current of 0.5A. What voltage is now needed?

Solution

Current $= 0.5$

Total resistance $= 5 + 2 = 7$

Voltage now needed $= 0.5 \times 7 = 3.5$ volts

Gear ratio

When two gears of different sizes are meshed together, the smallest gear will rotate much faster than the larger gear. The difference in speed of both gears is known as the gear ratio.

$$\text{Gear ratio} = \frac{\text{Number of teeth in driven gear}}{\text{Number of teeth in driver gear}}$$

MATHEMATICAL UNDERSTANDING
Gear ratio example

An input gear has 11 teeth and the driven gear has 19 teeth. Calculate the gear ratio.

Solution

Driver $\rightarrow$ input gear $= 11$

Driven $\rightarrow$ idler gear $= 19$

$$\text{Gear ratio} = \frac{\text{Number of teeth in driven gear}}{\text{Number of teeth in driver gear}} = \frac{19}{11} = 1.7$$

Gear ratio $= 1.7$ (to 1 d.p.)

This means that for every one turn of the driver (input) the driven (output) gear turns 1.7 times.

Mechanical advantage

Mechanisms are used in engineering to make a job easier to accomplish. The main types of mechanisms are levers and gear ratios. To calculate how much easier a job will be (mechanical advantage) by using a mechanism we use the formula below.

Mechanical advantage $= \dfrac{\text{Load}}{\text{Effort}}$ ← (Object to be moved)
    ← (Force needed to move the object)

MATHEMATICAL UNDERSTANDING
Mechanical advantage example

A prising bar is used to extract a bolt from a metal plate. The force applied to the prising bar is 40 N and the prising bar applies a force of 640 N to the bolt. Calculate the mechanical advantage.

Solution

Mechanical advantage $= \dfrac{\text{Load}}{\text{Effort}}$

$= \dfrac{640 \text{ N}}{40 \text{ N}} = 16$

KEY POINTS

Table 6.6.2 **Key equations**

Calculation	
Stress	$\dfrac{Force}{Cross\text{-}sectional\ area}$
Strain	$\dfrac{Change\ in\ length}{Original\ length}$
Density	$\dfrac{Mass}{Volume}$
Young's Modulus	$\dfrac{Stress}{Strain}$
Current	$\dfrac{Voltage}{Resistance}$
Gear Ratio	$\dfrac{Number\ of\ teeth\ in\ driven\ gear}{Number\ of\ teeth\ in\ driver\ gear}$
Mechanical advantage	$\dfrac{Load}{Effort}$

Select and use materials, parts, components, tools and equipment

What will I learn?

By the end of this chapter you should have developed knowledge and understanding of:
→ the factors to consider when selecting materials
→ the safe use of parts in manufacture
→ the safe use of common components found in engineering
→ selecting and safely using tools and equipment

When manufacturing a product, it is important to consider and select the correct materials, tools and equipment, which can be safely used. In some instances, parts and components will need to be chosen and used to complete the manufactured product.

Material selection

Material selection plays a key role in manufacturing a **working solution** as it determines the quality and reliability of the manufactured product. When selecting a material for a product, engineers look at the characteristics of varied materials – for example, aluminium is used for making cooking pans as it has good resistance to corrosion, is low density, is a good conductor of heat and can be polished to give a reflective finish. By assessing the material's characteristics in this way, you can decide whether it is suitable for its intended working conditions. For example, a metric spanner comes into contact with moisture and other liquids so it needs to have a corrosion-resistance finish, and it is used to tighten and slacken nuts so it needs to have high strength so it does not break.

> ### KEY WORD
>
> **Working solution:** a component or product that has been made to bring an idea or concept into reality.

Factors to consider when selecting materials

When selecting which materials are suitable for use, you should consider the following factors:
- Properties of materials – you should know the working properties of the material in relation to its application.
- Intended lifespan – it is important to know how long the product is intended to be in use for.
- Availability of the material – consider if the material is readily available (can the material be sourced locally or will it need to be transported in?) or if it has to be specially produced.
- Cost of the materials – consider the cost of sourcing (finding a supplier who can provide the needed items).
- Wear conditions – consider if moving parts will wear during use. Materials must be selected which have good wear resistance or can be heat treated to avoid wear. For example, screwdriver blades are hardened and tempered to avoid wear and breakage during use.
- Corrosion – the environment in which the materials will be used will need to be considered so they do not corrode in service.
- Manufacturing considerations – consider if the material is suited to the scale of production required to manufacture the product. For example, a high volume of parts and components are press formed as this cost-effective process means a high volume of parts can be made in a short amount of time.

Parts selection

A **part** is a pre-made article that is assembled to make a product. When selecting parts to make a product, the same factors as selecting materials apply (see above). The motor vehicle sector is a good example of manufacturing which uses parts to manufacture a product. Examples of parts used in the motor vehicle sector include body panels, brakes, fuel pumps, gear boxes, headlights, etc.

Figure 6.7.1 **A headlight part being fitted to a car**

Components selection

Standard components (for example, nuts and bolts, bearings and bushes, chains and sprockets, and screws and rivets) are mainly used to assist in the manufacture of a product. They come mainly in standard form – a nut and bolt is a standard form component as it is a mass-produced item and comes in a standard size which makes it cost effective and readily available. However, in some cases, standard components cannot be used for some applications. Specialised components may need to be made, which can be costly for the manufacturer as they have to be made in small batches rather than mass produced.

Figure 6.7.2 **Collection of set screws, nuts and washers**

Tools and equipment selection

Before manufacture can commence, it is advisable to produce a **manufacture working schedule** explaining the process stages of manufacture, and identifying the tools and equipment required to carry out each process so that the highest quality can be achieved.

Factors to consider when selecting tools and equipment

When selecting which tools and equipment are suitable for use, you should consider the following factors:

- the manufacture specification of the artefact
- materials being used to manufacture the product
- manufacturing tolerances
- the quantity to be produced.

Safe use of materials, parts, components, tools and equipment

It is very important that safety is considered when working with tools and equipment to safeguard every worker involved in the manufacturing process. Protective equipment and clothing should be worn (such as safety glasses) and safety precautions followed to avoid unnecessary accidents. Safety awareness should be included in the manufacture working schedule.

KEY POINTS
- Material selection is an important consideration in manufacturing a working solution to ensure the quality and reliability of the manufactured product.
- Standard components are used to assist with the manufacture of a product as they are cost effective and readily available.
- Before manufacture begins, a manufacture working schedule should be drawn up which outlines the process stages of manufacture and identifies the tools and equipment required.
- Safety at each stage of the manufacturing process should be considered so control measures can be put in place to safeguard the workers.

Check your knowledge and understanding

1 Describe four factors to consider when selecting materials for a product.
2 Explain why parts and components are used by manufacturers.
3 Name four different types of components.
4 Explain why a manufacturing working schedule is used in engineering.

6.8 Select and use appropriate processes

What will I learn?
By the end of this chapter you should have developed knowledge and understanding of:
→ how to select the appropriate processes that can be used to manufacture a product
→ how to correctly use the processes that can be used to manufacture a product.

When manufacturing a product, it is important to use the correct equipment and processes safely and correctly. Before manufacture begins, it is useful to produce a manufacturing working schedule so that processes can be carried out in the correct order.

Below is a list of different processes that can be used to manufacture a working solution.

Measuring

Measuring equipment is used to measure lengths and other geometrical parameters, such as length, diameter and depth. They can also assist with marking out. Different types of measuring equipment include:

Measuring with an engineer's rule
The engineer's rule is the most commonly used piece of measuring equipment. It is useful for quick measurements that can be taken from the scale on the body of the rule.

Measuring with callipers
There are usually two types of callipers used: internal callipers and external callipers. They are used to measure internal and external size, for example, measuring the internal diameter of a bore. Once the measurement has been taken by the calliper, a ruler (a scaled measuring device) is used to take the measurement of the calliper to give a scaled reading.

Measuring with a micrometer
A micrometer is a scaled measuring device that comes in different sizes from 0–25 mm to 150–175mm. They are used to check small distances with extreme accuracy. They are more accurate than a vernier calliper.

Measuring with vernier callipers
Vernier callipers can be used to measure both the internal and external faces of an object. They can take accurate measurements with their internal and external jaws.

Figure 6.8.1 The internal and external jaws on a digital vernier callipers

Measuring with a feeler gauge

A feeler gauge is a set of finely-ground tool steel strips which progress in thickness in relation to sizes. They are used to measure fine gaps between material faces.

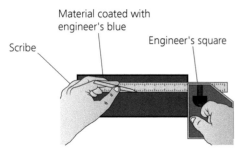

Figure 6.8.2 The process of marking out

Marking

Marking is the process of transferring measurements from a drawing or design to a workpiece. The workpiece material to be marked out is prepared by removing grease and oil with a degreasing solution. The material is then coated with engineer's marking paste – this is a blue pigment paste rubbed onto a surface which, when dry, leaves a blue film. This can then be marked out allowing the markings to stand out during manufacture. The most common tools used for marking out are an engineer's square, scribe, engineer's rule, dividers and odd-leg callipers.

Turning

Turning is a process used to produce rotational/cylindrical workpieces. The workpiece is rotated while the tool is stationary, unlike milling where the workpiece is stationary on the machine bed and the tool rotates. The different types of lathe turning operations include:

- Turning – the process of reducing the diameter of a cylindrical object. A taper and cones can also be produced by turning.
- Facing – the process of removing metal from the end of the workpiece so it is perpendicular to its length.
- Knurling – when a pattern is rolled onto the slowly revolving material using a knurling tool in the lathe tool post.

Figure 6.8.3 Turning process

- Screw cutting – when a thread is cut into the diameter of a workpiece using a single-point screw-cutting tool. The feed rate and rotation of the workpiece are set according to the pitch of the thread to be cut.
- Drilling – involves fitting the drill to the tailstock of the lathe using a chuck or a drill with a morse taper shaft. A hole is always started with a centre drill. A drill is used to drill a hole in the material being rotated in the chuck.
- Boring – the process of enlarging a drilled hole using a boring bar held in the tool post of the lathe.

The image shows the outer diameter of a cylindrical piece of work being turned down on a lathe.

Milling

Milling is the process of using a milling machine to remove material from a workpiece using a cutting tool which is rotating. The workpiece is stationary and is held on the table of the milling machine by a clamping device. The following are types of milling operations.

- Face milling – the process of machining flat surfaces or flat-bottom holes which are at right angles to the milling cutter.
- Milling a slot or recess – the same principle as face milling is used but a slot is cut into the material.

Standard types of milling cutters include end mills, slot drills, t-slot cutters, and ball nose slot drills.

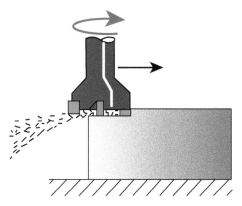

Figure 6.8.4 A face milling cutting direction

Figure 6.8.5 A slot being cut by a slot cutter milling tool

Figure 6.8.6 A workpiece being drilled in a machine vice

Drilling

Drilling is normally carried out using a pillar drill. A twist drill is placed in the chuck of the drill and tightened with a chuck key. The material is secured to the table of the drill directly or by an engineer's machine vice. Like milling, the cutting tool (the drill) is rotating and the workpiece is stationary.

Twist drills come in a range of different sizes from 0.2 mm to 25 mm. Cutting speeds are determined by the size of the drill being used. The smaller the drill, the faster the drill rotation. The speed of the drill can be adjusted on the pillar drill by adjusting the belts between the different size pulleys on the top of the drill.

Forming

No material is removed during the process of forming. The workpiece is reshaped using tools and different forming processes. Some types of forming operations include:

- Press forming – the process of forming material in a press. Material is formed into shape by a press, either by stamping or using a **die set**. Complex shapes can be produced quickly by using this process.
- Rolling – the process of passing the material through a set of rollers to reduce the thickness of the material, thus increasing the width of the material.
- Forging – the process where the material is heated, usually in a furnace or forge. The material is formed into shape with blows from a hammer.

Bending

Bending is most commonly used in sheet metal work to **fabricate** products. Bending can take place by using sheet metal benders or engineering hand tools, such as folding bars and mallets (for more details on hand tools see Folding below).

Hand bending

The material is marked out and the bending locations identified. Different thicknesses of materials will need to have different bend allowances. When sheet metal is bent, it stretches so its length increases slightly, which can take it out of tolerance to a working drawing.

To avoid this, a calculation can be carried out according to the dimensions of the material, which allows for this increase of length. This is known as the bend allowance.

After the bend allowance has been marked onto the material, it is then clamped in a bench vice fitted with a bend former, making sure it is lined up correctly to the marked line. A bend former has the shape of the bend machined into the edge of it so it contains the material while bending, allowing a perfectly shaped bend form. To prevent marking of the material, a rubber, copper or rawhide mallet should be used to lightly hammer the material over the bend former, making sure it is uniformly struck across the whole length of the bending seam or line until the desired angle or profile is obtained.

Figure 6.8.7 A hand-operated sheet metal bender

Sheet metal bender

The most common types of sheet metal benders used in schools are bench mounted and hand operated. The bending allowance is calculated and the material is then marked accordingly and placed into the bending machine. It is then clamped in place (bend lines in line with bending plate). The sheet metal bender is then hand operated until the required bend angle is produced. This can be checked with an engineer's protractor.

Casting

Casting is used to produce complex shapes out of metal or plastic. The molten material is poured into the cavity of the mould (the shape of the intended item). (See Section 2.4 Casting and moulding for more details.)

Joining

Joining is the process of assembling parts together. There are two joining methods: permanent joining and temporary joining.

Permanent joining methods

- Brazing – the process of welding two pieces of metal together by melting an alloy filler to bond the two pieces together. The filler has a lower melting point than the two metals that are being welded. Normally, a flux is used to clean the oxides of the metal away so the filler can flow and bond more tightly. It is used for joining metals such as silver, copper and gold. Brazing is not as strong as welding.
- Welding – this process uses very high temperatures to melt and join metal parts together. A filler metal is normally used to help with the joining process. Welding can be used to join similar metals, such as welding two pieces of low-carbon mild steel used in a lot of different engineering applications such as shipbuilding, car manufacture, jigs and fixtures manufacture.
- Soldering – this process is very similar to brazing, but a much lower temperature is required. Fillers (solder) are used to join the metal parts. Flux is used to clean the surfaces and make the solder flow. Solder is used to join electrical components.
- Rivets – a rivet can be used to join metal sheeting together. The most common form of riveting found in schools is the snap head rivet. Sheeting is overlapped and the rivet holes marked and centre punched. The sheeting is then drilled with a drill to the same diameter as the rivet. The rivet is then pushed through the hole of both sheets of metal and a rivet

gun is used to expand the rivet in the metal sheets so they join together. When the rivet reaches its maximum expansion size the head of the rivet snaps off leaving the formed rivet in the metal. This technique is used mainly for joining thin sheets of metal together where brazing or welding would damage the material because of its thickness.

- Metal glue – epoxy resin is a two part adhesive that can be used to join metals together. They are not as strong as welding but they do allow slight movement when set. They can be used in many different applications to join materials together.

When the load is high on applications, permanent joining materials are used.

Temporary joining methods

Temporary joining methods include:
- Nuts and bolts – these components are used to join unthreaded parts together. They are joined together by an internal thread on the nut and an external thread on the bolt and screw together to form a tight bond. Nuts and bolts come in different metric and imperial sizes.
- Machine screws – these use a pre-existing tapped hole to screw into. They come in a range of different head designs and sizes.
- Clamps – these have two clamping surfaces that can be positioned between an object. They are normally connected by a threaded bar. As it is tightened, it pulls the clamping surfaces together resulting in the object being clamped tightly.
- Metal hardened dowel pins – these are used to align parts together with precision. Each pin is machined with a slight interference fit for the hole they are to be located in, so they fit tightly and firmly. They are used in press tools to align parts together.

In most cases, temporary joining methods are easy to remove and disassemble parts.

Fastening

Fastening is a process used to hold two or more objects together. The joints are not permanent and so can be engaged or released when required. Types of fastenings include retaining rings, nuts and bolts, screw threads, taper dowel pins, cotter pins, and wing nuts.

Folding

Folding is the process of bending a material past 90° so it folds back on itself. This process is normally carried out on sheet metal. To fold sheet metal, the same process as bending is used, however, the material is bent more than 90° so folding bars can be inserted in the fold. The folding bars are struck with a hide mallet; in turn the metal is folded on itself. This process can be used to join metal sheets together. A wire can also be inserted to strengthen the edge of the fold.

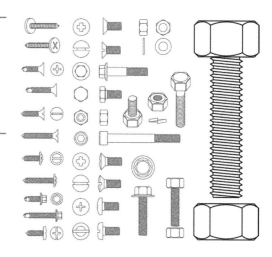

Figure 6.8.8 Different kinds of fastenings

Shaping

Materials can be manipulated into a shape using a number of methods. These methods can include using very small components through to large, complex castings.

Engineer's files can be used in the workshop to shape metals, alloys and plastics. The files cover a range of different cutting lengths.

Figure 6.8.9 A folded joint to hold two pieces of sheet metal together

Shaping

The types of cut produced by the shaping process include:
- Bastard – a coarse cut file for removing large amounts of material at one time.
- Second – less coarse than the bastard file. This is the most commonly used in the workshop.
- Smoothing – as the name suggests, this file is used to finish off or smooth areas that have been filed. They can be used to smooth rough edges after manufacture has taken place.

The types of files used for the shaping process include:
- Flat – used for filing flat or angled surfaces.
- Half round – used for filing flat or round surfaces.
- Round – used for filing a round profile and enlarging holes. It can also be used for filing tight curves.
- Square – used for filing on narrow flat surfaces, it can also be used to enlarge holes and slots.

Machinery such as milling machines are more commonly used to shape material.

Finishing

Materials will require finishing to improve their look as well as to protect them from their working environment.

Surface finishes are applied to materials to:
- protect the material from abrasion, corrosion, wear and moisture
- improve the material's durability and surface hardness
- enhance the aesthetic properties of the material.

Different types of finishes that could be applied include:

- Painting – the material is degreased and a base coat (undercoat) is applied to the material, either with a brush or by spraying. The base coat is allowed to dry before repainting with the final colour in either gloss or matt paint, depending on the desired effect. To obtain the best finish, all holes and dents need to be filled and prepared prior to painting.
- Dip coating – the material is coated with a thin layer of polyethylene plastic to protect it from corrosion. The painting process is carried out by heating the metal to a cherry red, then placing it into a fluidised bath of polyethylene powder, where air is circulated around the bath to make the powder airborne. The powder sticks to the heated metal uniformly and solidifies to a shiny plastic coating when cooled. Different coloured powders can be used to produce different coloured finishes.
- Electroplating – a process used to electroplate metals. Expensive metals such as silver can be used to plate low-carbon mild steel. Cutlery is one of the most popular electroplated products. The electroplating process is carried out by connecting the piece of cutlery to be plated to the negative terminal of a power supply. A piece of silver is connected to the positive terminal of a power supply. Both terminals are submersed in a solution of silver nitrate solution (this is known as the electrolyte). As current is passed through the circuit the silver gradually moves from the positive terminal to the negative terminal where it is deposited onto the cutlery (plating the cutlery with silver).
- Anodising – a process, usually using aluminium, to provide an additional corrosion-resistant surface. It increases the thickness of the surface, making it more durable and with a better resistance to wear. The plating process involves electrolysis in an acid bath solution.
- Galvanising – a process using molten zinc to plate metals such as low-carbon mild steel to protect them from corrosion. The plating process is carried out by firstly cleaning and degreasing the metal to be plated, then dipping into a bath of molten zinc. The zinc bonds to the steel. When cooled, it forms a hard layer of zinc which is corrosion resistant.
- Oil blackening – the process of heating metal to a dull red colour and then quenching (rapidly cooling down) in used engine oil.

KEY POINTS

- You must use the correct equipment and processes safely and correctly when producing products.
- Measuring equipment is used to measure lengths and other geometrical parameters such as diameter and depth.
- Marking involves transferring measurements from a drawing or design to a workpiece.
- Joining (both permanent and temporary) is the process of assembling parts together.
- Materials can be finished to improve their look and to protect them from their working environment.

Check your knowledge and understanding

1 Name three different types of measuring equipment.
2 Explain why engineer's blue is used when marking out.
3 Identify which milling operation and cutter would be used to cut a flat-bottom hole in a piece of low-carbon mild steel.
4 Explain the difference between temporary and permanent joining methods.
5 Discuss the benefits of applying surface finishes to materials.

Apply quality control methods and techniques

What will I learn?

By the end of this chapter you should have developed knowledge and understanding of:

- why quality control methods and techniques are used in engineering
- the importance of working to necessary tolerances and the tools used to check tolerances
- how to use CNC/CAM software to ensure parts and components fit and operate correctly.

As explored in Section 4.2 Testing (see pages 124–131), **quality control** methods ensure that products are made to the highest standard achievable. They do this by checking/measuring products at various stages of production. The information is recorded, and any necessary adjustments are made to maintain the high standards of production.

Refer back to Section 4.2 Testing for more information on the different test types that are used in engineering processes.

Working to necessary tolerances

Each quality control inspection stage has set tolerances that must be used to quality check a product to see if it can be accepted. The measurement must fit within the set limit (in tolerance) or out of the set limit (out of tolerance). The sample is then rejected or reworked.

Factors that could affect the size of products during manufacture

During manufacture, most parts are within tolerance. However, if measured against the original drawing dimensions (which is called the nominal size), there is a high risk that they will not match the dimensions exactly. On large manufacturing runs, it is very difficult for the products to be made to the exact size of the dimensions on the drawing, so go/no-go gauges are set to allow this slight difference in size. As long as they are within the go/no-go gauge size (this is known as the agreed/**acceptable tolerance**) they will be fit for purpose and can be used.

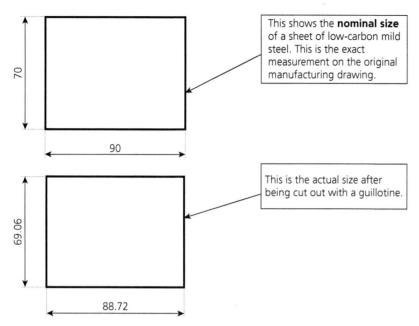

This shows the **nominal size** of a sheet of low-carbon mild steel. This is the exact measurement on the original manufacturing drawing.

This is the actual size after being cut out with a guillotine.

Figure 6.9.1 Dimensions of two sheets of low-carbon mild steel.

The acceptable tolerance is 90 ± 2.0 × 70 ± 2.0 mm.

The following are reasons why product sizes can change during manufacture:
- The cutting tools can become blunt and worn from constant use.
- Cutting tools can move during operation from vibration of machinery.
- Imperfections in materials can affect the cutting action of the tools.
- Improper setting up of machinery can cause cutting tools to move during operation.
- Lack of lubrication/coolant can cause cutting tools to overheat and expand causing excessive cutting of materials.

Quality control techniques

Many products are made up of different components which are made in different areas of a production plant or elsewhere by a different company. To ensure they fit together at the final assembly stage, regular quality control checks need to be put in place. This usually occurs after each machining process. The check ensures the components are within tolerance.

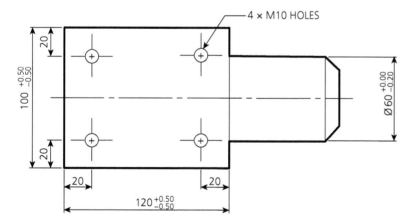

Figure 6.9.2 A pivot block with tolerances for manufacture

Table 6.9.1 The quality control checks that could be done on a component at different machining processes

Machining process	Quality control check	Tools to check tolerances in school	Go/no-go gauge to check tolerances in industry
Machine blank to size	Has the blank been cut out to the correct size?	Vernier calliper Micrometer	Height gauge
Drill 4 x 10 mm diameter holes through blank	Are the holes in the correct location?	Vernier calliper	Positioning gauge – to check location
	Are the holes the correct size?	Vernier calliper	Plug gauge – to check hole size
Tap 4 x M10 holes	Are holes tapped correctly?	Vernier calliper / 10 mm	Thread gauge
Turn blank to required diameter	Has the blank been turned to correct size.	Micrometer to check diameter	Snap gauge to check diameter
		Vernier callipers to check depth	Height gauge to check depth

Tools to check tolerances

These are tools that can be used to check a workpiece against its upper and lower tolerances.

Go/no-go gauge

A go/no-go gauge is a very fast and effective way of checking a component as it does not have any measurement scales to read. The gauge has two check areas or test areas:

- Go – for components that are in tolerance
- No-go – for components that are out of tolerance.

Types of go/no-go gauges include:

- Snap gauge – for checking outside diameters.
- Plug gauge – for checking internal diameters.
- Thread snap gauge – for checking threads (pitch of thread).
- Depth gauge – for checking depths of bores.
- Height gauge – for checking set distances.

Other tools to check tolerances include the vernier calliper and the micrometer. Although accurate, they are not as effective as a go/no-go gauge, which is regularly checked over and calibrated (adjusted) during production runs to ensure quality assurance.

Using software (CNC/CAM) to ensure that all parts/components fit together

The CAD software helps the designer check the final concept/product by bringing all parts together in a final assembly. Animation can then be applied to the assembly to see if it works. Modifications can easily be made if required. The assembly can also be tested in different working conditions to allow materials to be trialled for performance and suitability.

When the designer is happy with the final assembly, a working model can be machined using either CAM, the intended material, or by using additive manufacture (3D printing). After model trialling has taken place, any modifications are fed back to the designer, who makes necessary modifications. Tolerances are then applied to parts before they go into continuous production.

KEY POINTS

- Quality control ensures that products are made to the highest standard achievable.
- Each quality control inspection stage has set tolerances that must be used to quality check a product.
- A go/no-go gauge is a quick and reliable means of checking a component as it does not have any measurement scales to read.
- CAD software helps the designer check the final product by bringing parts together in a final assembly.
- Animation can be applied to the assembly to see if it works.

Check your knowledge and understanding

1 Describe why quality control is an important part of the manufacturing process.
2 Explain why it is necessary to work to tolerances when manufacturing products.
3 Give two reasons why go/no-go gauges are so widely used in engineering.
4 Suggest a suitable use for a plug gauge.
5 Explain how CAD software could be used to ensure all parts and components of an assembly fit together correctly.

Design tests to assess fitness for purpose and performance

What will I learn?

By the end of this chapter you should have developed knowledge and understanding of:

→ different types of testing used to assess a product's fitness for purpose and performance
→ how to record test results
→ how to interpret test results to identify areas for improvement/modification
→ considering alternative solutions.

Once a product has been manufactured and released to the general public, it should be **fit for purpose**. A good way of assessing this is to design a range of tests that can be carried out on the product.

Once you have manufactured a product following your design and production plan, it should be fit for purpose. However, a good way of assessing this is to design a range of tests that can be carried out on the product.

Types of product tests

There are two types of tests that can be carried out to assess the fitness for purpose and performance of a completed product:

● **performance testing**
● **environmental testing**.

Performance testing

Performance testing is one of the most important testing areas as it tests if and to what extent the product performs as it was initially planned to do. A performance test is broken up into smaller tests which can be recorded and then a graph plotted to help understand the following questions about the product.

● How easy is it to use?
● Is it the correct weight for its application?
● Is it the correct size for its intended location?
● Does it function correctly?
● Does it do the job it was designed to do?

These types of tests are also known as **product testing/consumer testing** and are carried out by individual people who are not connected to the manufacturer. They usually have a questionnaire or table to record test results, similar to the one shown in Table 6.10.1 (page 196).

Environmental testing

Environmental testing is carried out to make sure the product has been manufactured with sustainability in mind – this means to reduce the negative impact on the environment as much as possible. The manufacturing history (series of manufacturing events) needs to be

Table 6.10.1 Test results for desk lamps: fitness for purpose and performance

Type of testing	Description of test	Lamp 1	Lamp 2	Lamp 3
Performance	Easy to operate	1 2 3 4 5 6 7 8 **9** 10	1 2 3 4 5 6 7 8 9 **10**	1 2 3 4 5 6 **7** 8 9 10
Performance	Easy to carry	1 2 3 4 **5** 6 7 8 9 10	1 2 3 4 5 6 7 8 **9** 10	1 2 3 4 5 6 7 **8** 9 10
Performance	Correct size for situation	1 2 3 4 **5** 6 7 8 9 10	1 2 3 4 5 6 7 8 **9** 10	1 2 3 4 5 6 7 8 **9** 10
Performance	Stable during operation	1 2 **3** 4 5 6 7 8 9 10	1 2 3 4 5 6 7 8 9 **10**	1 2 3 4 5 6 7 8 **9** 10
Performance	Brightness of light	1 2 3 4 5 6 7 8 9 **10**	1 2 3 4 5 **6** 7 8 9 10	1 2 3 4 5 6 7 8 9 **10**
Environmental	Use of locally sourced materials	1 **2** 3 4 5 6 7 8 9 10	1 2 **3** 4 5 6 7 8 9 10	1 2 3 4 5 6 **7** 8 9 10
Environmental	Sustainable manufacturing methods used	1 2 3 4 5 6 **7** 8 9 10	1 2 3 4 5 6 7 8 **9** 10	1 2 3 4 **5** 6 7 8 9 10
Environmental	Use of recycled materials	1 2 3 4 5 6 7 **8** 9 10	1 2 3 4 5 6 7 **8** 9 10	1 2 3 4 5 **6** 7 8 9 10
Corrosion	Resistance to salt water	1 2 3 4 5 **6** 7 8 9 10	1 2 3 4 5 6 7 8 **9** 10	1 2 3 4 5 6 7 8 **9** 10
Visual	Parts fitting together correctly	1 2 3 4 5 6 7 **8** 9 10	1 2 3 4 5 6 7 8 **9** 10	1 2 3 4 5 6 7 **8** 9 10
Visual	Overall finish of product	1 2 3 4 5 6 7 8 9 **10**	1 2 3 4 5 6 7 8 9 **10**	1 2 3 4 5 6 7 8 **9** 10
Quality	Are parts within tolerance?	1 2 3 4 5 6 **7** 8 9 10	1 2 3 4 5 6 7 8 **9** 10	1 2 3 4 5 6 **7** 8 9 10
Failure test	Operation of switch until failure	1 2 3 **4** 5 6 7 8 9 10	1 2 3 4 5 6 7 **8** 9 10	1 2 3 4 5 6 7 8 **9** 10
Failure test	Tilting lamp up and down until failure	1 2 3 4 **5** 6 7 8 9 10	1 2 3 4 5 6 7 8 9 **10**	1 2 3 4 5 6 7 8 9 **10**
Safety	Checking for heat conductivity	1 2 3 4 **5** 6 7 8 9 10	1 2 3 4 5 6 7 8 9 **10**	1 2 3 4 5 6 7 8 **9** 10
Safety	Checking for electrical insulation	1 2 3 4 5 6 7 8 9 **10**	1 2 3 4 5 6 7 8 9 **10**	1 2 3 4 5 6 7 8 9 **10**
	Total	104	139	132

Key: 1 = Very Poor; 10 = Excellent

traced to make sure that it has been produced with the fewest detrimental effects on the environment. Factors which can reduce the environmental impact of materials include:

- materials that have been sourced locally, to avoid pollution from transportation
- using recycled materials wherever possible for manufacture
- using non-toxic paints and packaging materials for packaging and finishing the product, which do not poison or damage the environment
- dismantling and recycling the product at the end of its working life without releasing harmful and toxic elements.

Product tests to identify areas for improvement

As covered earlier, other tests you can use to identify areas for modification and improvement, or consider alternative solutions are as follows:

Corrosion testing

Corrosion testing is commonly used by manufacturers to check how much resistance to corrosion the product has. Commonly, a mixture of salt and water is sprayed onto different areas of a product to see if it can withstand this corrosive substance. After a couple of days, the areas treated are checked for any forms of corrosion. This test can be carried out on different surface finishes to see what protection they offer to the underlying material.

Non-destructive testing – visual testing

Non-destructive testing – visual testing checks the visual quality of the product, or how 'aesthetically pleasing' (attractive) it is. It can also make sure that all parts are finished correctly and find faults such as missing pieces or parts not fitting correctly together.

Quality testing – accuracy

Quality testing – testing for accuracy, or how well the products matches the manufacturers' specification. It tests that each part of the product is within the manufacturing tolerance. Each part of the product has to be measured and results recorded.

Destructive testing – failure testing

Destructive testing – in failure testing, the product is tried in different working environments, with different loads/forces continually applied to the product until failure occurs. The time taken for this to occur is recorded. This test may take a couple of months to complete.

Safety testing – to British and European Standards

Safety tests help to predict and reduce possible hazards when using the product; for example, that electrical components have been fully insulated, folding parts cannot trap fingers, etc.

Tests carried out on products can be recorded in a table. Table 6.10.1 is an example of the testing that could be carried out on products such as lamps. The table is an easy way of recording and showing results, which can easily be referred to in the final testing and evaluation section of the NEA.

Results and suggested modifications

Once the product has been tested, it will be clear from the test results how different parts of the product have performed. Testing might reveal the product corrodes in its working environment. As a result, either the manufacture materials or type of surface finish needs to be reconsidered. If the electrical switch fails after being continually tested on and off for 300 hours, the manufacturer discovers a more substantial switch (better quality) is required. Sometimes, only part of a product may fail, which will need to be addressed by the manufacturer who will make necessary modifications to that part before the product is released to the general public.

ACTIVITY

Think about a product you have made in your engineering lessons. Produce and carry out a range of tests on this product, recording the results as you carry out each test.

From the results, make recommendations about how you could modify the product you have initially tested so it would perform better if the tests were to be carried out again.

KEY POINTS

- Performance testing tests if and to what extent the product performs as it was initially planned to do.
- Environmental testing is used to make sure the product has been manufactured with sustainability in mind.
- Product testing/consumer testing is carried out by individual people who are not connected to the manufacturer, to provide unbiased feedback.
- Modifications are carried out to improve the characteristics of a product after testing has taken place.

Check your knowledge and understanding

1 Explain why testing is carried out on products.
2 Describe how performance testing could be carried out on a product.
3 Explain how and why a corrosion test is carried out on a product.
4 Identify which modification could be carried out to a product if it scored very low in a failure test.

1 Explain **two** reasons why engineers analyse existing systems.

2 A tie bar is 2.5 metres long, wide and 6 cm × 19 cm high. A pulling force of 3000 KN is applied to the bar, resulting in a 2.5 cm change of length to the material. Find the:
   i   Stress
   ii Strain
   iii Young's Modulus of elasticity of the material.

3 A portable electrical appliance consumes a constant power of 4 kw when the supply is at 240V. Determine the resistance of the appliance in ohms.

4 The image below shows a simple gear train. What is the gear ratio of the gear train? Give your answer to 1 decimal place.

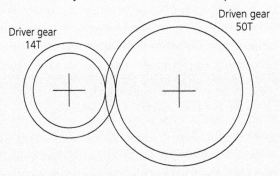

Driver gear
14T

Driven gear
50T

5 Complete the following statement using the word bank provided.

_____ (CNC) uses a binary number system called _____ to operate the _____ and _____ rate of the machine. CAD software can be used to produce a _____ for the computer _____ machine used to manufacture _____.

Word bank: program, feed, products, computer numerical control, aided, machine code, speed.

6 What are these types of lines used for in an orthographic drawing?
   i
   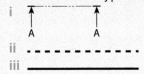
   ii
   iii

7 A hardness test is carried out on material to find the best …
   A finish to be used to protect the material.
   B machine tools to be used to machine the material.
   C best way to cast the material into a shape.

8 Which equipment could be used to mark out a shape into a sheet of material?
   A Vernier calliper
   B Scribe
   C Folding bar
   D Odd-leg callipers
   E Callipers
   F Boring bar

7 Preparing for assessment

You will be formally assessed on how well you have understood the content of the course. It is important that you understand what is involved.

You will sit a written examination paper and submit a project known as the NEA (non-exam assessment). The written paper will be worth 60 per cent of your final grade, whereas the NEA is worth 40 per cent.

This section explores each of these two assessments in detail and gives tips on how to attempt them successfully.

This section includes the following topics:

7.1   The written paper

7.2   Non-exam assessment: practical engineering

7.1 The written paper

What will I learn?

By the end of this section you should have developed knowledge and understanding of:

→ when you will complete the written examination
→ how long you will have to complete the written examination
→ the basic format of the paper and the question types you may encounter
→ the criteria that you will be assessed against in the written exam.

The exam paper is worth 60 per cent of the assessment for your AQA GCSE Engineering qualification. It is important to ensure that you are well prepared for the types of questions that you may be asked.

When will I take the written paper?

The written paper will take place in the summer exam period. Your school or college will inform you of the exact date and time of the exam as this will vary from year to year. Alternatively, you can check for yourself via the AQA website: www.aqa.org.uk.

How long will I have to complete the exam?

You will have two hours to complete your examination and you should use all of the time available to you.

You will gain a better understanding of the time needed to complete the questions by practising them under exam-style conditions. This will show you which types of questions take you more time. You can plan how you will tackle the paper accordingly, as well as knowing where you need to spend more time practising before your examination.

Figure 7.1.1

What format will the written paper take?

The total number of marks available in the written paper is 120. There will be a mixture of different question types that you will be required to answer including:

- multiple-choice
- short-answer questions
- extended-response questions.

These will be explained in more detail below, starting on page 183.

Command words

When answering exam questions, it is very important to carefully consider what they are asking you to do. It helps to familiarise yourself with the way questions are worded to help you respond in the correct way. Taking note of the command words used in each question is a useful approach as these will indicate to you what the question is essentially asking you to do. You will then know how to construct your response. Common examples of command words include:

- calculate
- state
- describe
- explain
- discuss
- analyse
- evaluate.

Others may also be used; a full list can be found on the AQA website.

Calculate

'Calculate' means that you must work something out numerically. You might be asked to make calculations to check your mathematical understanding in any of the three question types you are required to answer. You should show all formulae and working that enable you to come to the correct answer. You should present all of your answers using the correct units and simplify where appropriate.

> Two resistors are connected together in series. They have values of $1k\Omega$ and $2.7k\Omega$.
> Calculate the total resistance. **[3 marks]**

When tackling this question, you should begin by writing down the correct formula that you are going to use, which is $R_t = R_1 + R_2$. You should then place the values for R_1 and R_2 into the formula and add them up. Finally, you should write down the answer including units, which in this case would be $3.7\ k\Omega$.

You will be expected to remember any formulae that you may need to use. There will be no formula sheet given in the examination.

State

Questions using the command word 'state' assess basic knowledge recall. They require a simple statement or short answer.

State the formula for Ohm's Law. **[1 mark]**

For this question you simply need to write down the correct formula, $V = I \times R$. You are not being asked to describe or explain your answer, so do not waste time adding any further information about voltage, current, etc. unless the question specifically requests this. Mistakes on this question would typically include writing down the wrong formula, or putting V, I and R in the wrong order within it.

Describe and explain

Questions that ask you to 'describe' are looking for a factual account of a feature, a characteristic or a process. For example, describing the steps taken to manufacture an engineered product.

An 'explanation' should go beyond a descriptive response. For example, it should include justifications for a particular choice and/or reasons for something occurring.

Explain two advantages of using a gear train instead of a pulley system to transmit drive. **[4 marks]**

In this question, you are being asked to explain two advantages, so your answer must go beyond two simple statements.

Typical answers could refer to:
- potential slipping of the belt on a pulley system, leading to breakage of the belt
- gear trains requiring far less maintenance, resulting in reduced cost.

Common errors in this type of question include not explaining your answer sufficiently or only writing about a single advantage instead of the two requested. You are not being asked to write about disadvantages, so only add them if the question specifically asks for them.

Discuss, analyse and evaluate

Questions which use 'discuss', 'analyse' and 'evaluate' require you to demonstrate your in-depth understanding of a topic. A discussion should cover all potential sides of an argument before coming to a balanced and appropriate conclusion. An analysis should break down the content to show understanding of it. An evaluation should look at several options or ideas and assess their success or worth.

Discuss the statement 'New and emerging technologies have a positive impact on the environment'. **[6 marks]**

This question is asking for a discussion of the statement presented, so you should aim to consider both positive and negative impacts before coming to a balanced conclusion. You should add detail to your points made and discuss them in depth. Typical answers could refer to:
- The use of electric engines to reduce carbon emissions from road and rail vehicles.
- The use of smartphone technology to control and reduce energy usage in the home. A counter argument to this could be that producing and transporting these products to the point of sale also creates carbon emissions. E-waste is created on disposal.

- The use of fuel cells to power buildings and transport – the only by-product of these is water.
- A conclusion 'summing up' the points made and coming to an overall judgement about whether or not new technologies have a positive impact on the environment.

A common mistake on this type of question is presenting a list of basic points rather than a discussion. In addition, a conclusion may also be absent or inappropriate. Make sure your conclusion takes into account the points made and is justified.

Figure 7.1.2

Multiple-choice questions

Multiple-choice questions will assess a breadth of knowledge from across the specification. They could also be related to the application of practical engineering skills. For each multiple-choice question, you will be given several choices and you must select the correct answer(s) from these choices. Sometimes there may be more than one correct answer that should be selected, so you should aim to read the question very carefully. Below is an example of a multiple-choice question. You should fill in the space in the box (it is known as a lozenge) of the correct answer.

Which of the following is a renewable source of energy?	*[1 mark]*
A Coal	☐
B Gas	☐
C Oil	☐
D Solar	☐

The correct answer from the options above is D – solar. All the other sources of energy shown are non-renewable. Other examples of renewable energy sources that could appear in the question are wind, tidal and biomass. When answering this type of question, double check you have filled in the correct lozenge. You do not need to write any additional information.

Short-answer questions

Short-answer questions assess your knowledge and understanding in more depth. They usually require a statement, a short descriptive response or an explanation. Short-answer questions will typically be worth between one and four marks.

Describe the injection moulding process.	*[4 marks]*

This question is asking the candidate to describe a process. It should read as a set of descriptive steps or points that would ensure that the injection moulding process is completed successfully. For this example question, answers could refer to:
- pouring plastic granules into a hopper
- heating the tube/barrel to a high temperature
- melting the granules into a liquid via a reciprocating screw
- feeding the melted plastic into a mould to form the shape.

Common errors in this type of question could include not describing each stage in enough detail, or missing out vital parts of the process. Make sure you include everything that is relevant in your answer but do not waste time going beyond what is being asked for – so, in this example question, you are not being asked to write about the benefits of the process. You should also not waste time writing about other processes that you have not been asked about – you won't gain any extra credit for doing this!

Extended-response questions

The third type of question that you could be asked are extended-response questions. These are questions that will bring together your knowledge of all the topics you have learnt about on this course. They will really test your depth of knowledge and will require you to write a broad response which covers a number of points in depth. You may be required to analyse and evaluate to present a really well-reasoned argument.

You may wish to plan out your answer to these types of questions before you begin writing them. For example, you could create a mind map to help you explore your ideas for your response. This involves noting down all your ideas, so you can organise them into a useful structure before you start writing.

You may be asked to include drawings, notes and/or sketches in your answer. If this is the case, try to draw and label them as clearly as you can.

These types of questions are usually worth six or more marks.

> Engineering industries can have both positive and negative impacts on society. Using at least two examples of engineering industries, evaluate the impacts that they have had on society. *[8 marks]*

This question is asking for a detailed evaluation that covers the topic in depth. It should go beyond a basic description or list of statements. To show your understanding, you should extend each point made and consider its importance. You should come to an appropriate conclusion. Typical answers to this example question could refer to:

- Changing behaviours on a mass scale, such as the mobile industry impacting on how people communicate with each other. Is this a good or a bad thing?
- The impact of different engineering businesses on employment and the consequences for people both locally and nationally. For example, a manufacturing facility opening nearby could bring extra jobs but also increase pollution, potentially putting extra stress on local healthcare systems.
- Improvements to transport infrastructure, such as better rail links that improve journey times to and from work.
- A conclusion summing up the points made and coming to an overall judgement about whether or not engineering industries have had a net benefit to society.

You should cover both positive and negative impacts in depth. As examples are being asked for, you should aim to include them and show how they are relevant to your answer.

How will I be assessed?

The written exam paper will assess you on the six topic areas which have been covered in this book:

1 Engineering materials (see Section 1)
2 Engineering manufacturing processes (see Section 2)
3 Systems (see Section 3)
4 Testing and investigation (see Section 4)
5 The impact of modern technologies (see Section 5)
6 Practical engineering skills (see Section 6)

The practical engineering skills section will be largely covered by the non-examination assessment (NEA) (see Section 7.2 below), but you will still be expected to answer questions related to practical contexts in this written paper.

There are three assessment objectives (AOs) that you will be assessed against in this written examination.

- AO1 will assess your ability to demonstrate knowledge and understanding of engineering principles and processes.
- AO2 will assess your ability to apply knowledge, understanding and skills in different contexts. This will include the use of a range of tools, equipment, materials, components and manufacturing processes.
- AO3 will assess your ability to analyse and evaluate evidence in relation to a range of engineering contexts.

Twenty percent of the marks will be allocated to questions that assess your ability to apply engineering-related maths skills.

General advice on answering exam questions

It is important that you are prepared for your written examination. You should aim to understand the different types of questions that you will be asked and what you are expected to do in order to answer them successfully. You can also make use of the specimen assessment materials available from the AQA website, which will give you an idea of what your paper will look like and what is expected of you in the allotted time.

If you have any spare time at the end of the exam, you could go back and check through your answers for errors and attempt any questions you may have missed out. You could also use this time to think about how you could improve existing answers or add any extra information.

7.2 Non-exam assessment: practical engineering

What will I learn?

By the end of this section you should have developed knowledge and understanding of:

→ the purpose of the non-exam assessment (NEA) and when you will complete it
→ how long you should spend on the NEA
→ the basic format of the NEA
→ the criteria that you will be assessed against
→ how to successfully complete each section of the NEA.

The non-exam assessment, or NEA, accounts for 40 per cent of the marks for the AQA Engineering GCSE. It will take the form of a 'design and make' activity based on a brief set by the AQA examination board. This chapter outlines the main things that you need to be aware of when completing this task.

Throughout this chapter examples of students' work are shown that reflect some of the assessment criteria. Explanations of these are given throughout the chapter to help you when preparing your own work.

When and where will the NEA be completed?

AQA will release the NEA brief for the task that you must complete, on 1st June in the academic year before you will be required to submit your work. This will then be passed onto you by your school or college. The context for this will be broad but AQA will provide you with three examples of how you might approach it. Over the following few months, you will work on your task, completing everything in time for your mark to be sent to the examination board. Over the following few months, you will work on developing and producing your engineered product.

The majority of the work should be completed under supervision from your teacher. You may take some work home but make sure that you check with your teacher before you do so. Remember that all of the work should be your own – If your teacher is unsure that the work has been completed by you, they cannot give you credit for it.

> ### NEA CONTEXT AND PROBLEM
>
> **The 'context' provides you with an introduction to the background of the engineering problem that you are going to solve, along with why it is important that it is tackled. The specific 'problem' will then be outlined. Three examples of possible solutions will be provided to help you.**
>
> **An example of this could be the potential for disasters at sea. This could result in possible scenarios such as oil tanker leakage, fire on a rig, people needing rescue, locating wreckage/aircraft black boxes or the use of warning beacons.**
>
> **A sample candidate booklet for the NEA, which includes an example context and problem statement, can be found via the AQA website.**

How long should I spend on the NEA?

You should spend around 30 hours completing the task. Your teacher will give you deadlines and target dates to help you manage your time and workload as this may be the first large project you have worked on. However, it is your responsibility to ensure that everything is done on time, so make sure you plan out all the required tasks in advance and stick to the timings set.

You will need to have everything done in good time so that your work can be marked, and your results sent to the exam board.

What format will the NEA take?

You are required to produce a working prototype of an engineered product that satisfies the brief. You may use a range of appropriate tools and equipment to do this (return to Section 2, Engineering manufacturing processes). Your teacher will show you what resources are available in your school or college and provide access to them. However, it will be down to you to select what tools and equipment would be the most appropriate for use in creating your 'solution' (your end engineered product).

Alongside your product, there should be a design portfolio of evidence containing:
● investigations into the context
● an analysis of the problem
● relevant research conducted
● designs for your solution
● a test plan
● a final evaluation of your prototype.

You should clearly show your methods for solving the design problem through the use of modelling (see Section 4), systems diagrams and relevant engineering drawings (see Section 3).

You may present the design portfolio in digital form if you wish. If you choose this option it must all be contained on a single file, such as a PowerPoint, Word or PDF file. Alternatively, a written portfolio is also acceptable. Photographs of your practical work should be taken throughout all stages of your design and development and added to this document as evidence that you have produced an engineered product.

How will I be assessed?

You will be assessed against a set of marking criteria split into six different sections. The breakdown of the marks available for each of these is shown below and explained in further detail in the remainder of this chapter.

Marking criteria/section	Marks available
Problem solving	15
Drawings and conventions	15
Production planning	15
Engineering skills used	15
Applying systems technology	10
Testing and evaluation	10
Total	**80**

Problem solving (15 marks)

In this section, you will need to:
- analyse the problem
- narrow down the context to a single problem worthy of further investigation
- come up with potential solutions
- model your ideas using a range of techniques
- clearly communicate and sketch these potential solutions
- produce a final prototype.

You will need to analyse (examine in detail) the problem as thoroughly as you can. This could involve using mind maps to explore and organise your thoughts and ideas. You should then ensure that you present a thorough written description of the task that you are going to complete. You should aim to clearly define the engineering problem that you are going to solve. For example, producing a prototype of a lifting device that can be deployed from above, to help people trapped on a boat at sea.

Try to organise and communicate your ideas clearly, providing explanations to justify your choices for how you will solve the problem. For example, label any sketches of product ideas that you produce and use notes to explain what they are and how they would work. This should include appropriate modelling techniques, such as 3D, graphical and mathematical. You can read more about doing this in Section 6, Practical engineering skills.

You can demonstrate your making skills by producing components using as many processes and as wide a range of skills as possible.

Do not worry too much about formal engineering drawings at this stage, as these will be considered later.

As explained above, when modelling ideas and producing a prototype, remember to take photographs to evidence your work in progress and the final product.

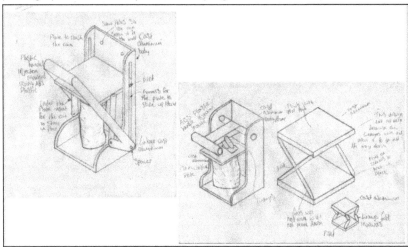

Figure 7.2.1 **Initial sketches for a can crusher design**

The sketches above show ideas for a can crusher product. Labels and notes have been used to explain the main features of the design and how it would solve the original problem – i.e. developing a product to aid in the recycling of waste. Sketches such as this are very useful for showing early ideas that can then be developed into a final design.

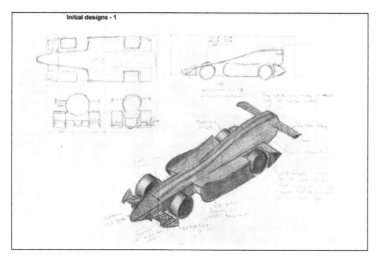

Figure 7.2.2 **Initial sketches for a racing car design**

The sketches above show an idea for a racing car. As with the example above, notes and labels have been added to explain the idea. This could be in response to a context that focuses on racing or increasing the speed of vehicles.

Drawings and conventions (15 marks)

For this section, you will need to:
● provide detailed, annotated drawings for your design idea which
  – use appropriate formal engineering drawings
  – comply with sector-specific standards and conventions
  – make use of CAD drawings for presentation.

You will need to demonstrate your ability to explain and communicate your design ideas using formal engineering drawings. These drawings should be appropriate to the problem that you are trying to solve. Examples of the types of drawings that you could produce include orthographic and sectional views, isometric drawings, assembly drawings and circuit schematics. All drawings produced should aim to use accepted engineering standards and conventions.

Computer aided design (CAD) software should also be used where appropriate, for example, to produce drawings of complex parts and 3D presentations. You should annotate your drawings in detail so that others can understand them. This could include notes on how the ideas will work, what materials/components have been used and how they will be manufactured.

Figure 7.2.3 shows how all the different parts of a racing car would fit together. This would be a useful type of drawing to include in response to most engineering problems presented in the NEA, as it clearly shows the location and purpose of each individual part of the proposed product. Explanations have been added to clarify the purpose and function of each part.

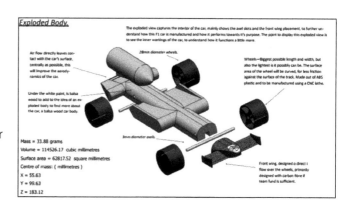

Figure 7.2.3 **An exploded view of a racing car design showing how the different parts are assembled**

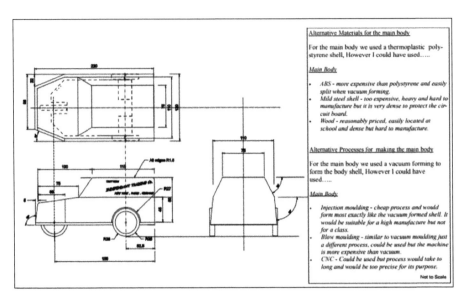

Figure 7.2.4 Orthographic drawing for the main body of an electronic buggy

This is an orthographic drawing for the main body of a moving electronic buggy. Notes have been added to explain how it might be manufactured, and different alternatives for this have been considered. This type of drawing is useful as it shows the different sides of your proposed design by using three different views. Think about how you could use orthographic drawings for the task presented in your NEA.

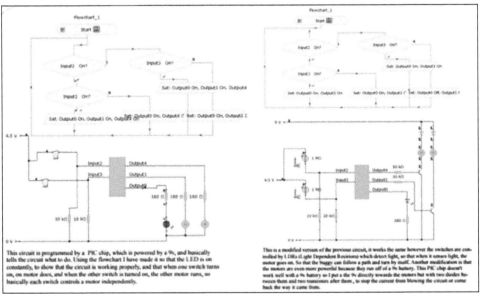

Figure 7.2.5 Circuit schematics and flowchart programs for a microcontroller-based electronic buggy

Shown above are examples of circuit schematics and flowcharts for a microcontroller-based system. If you are to use electronic or programmable systems in your NEA, then this is one good way to communicate them. You should aim to include all steps needed for any programs written and use the correct symbols in any circuit diagrams. Notes and labels can help to explain how the system would work. If you use circuit design software you can also simulate how the system would work before making it.

Production planning (15 marks)

For this section, you will need to:

● produce and follow a detailed production plan explaining each stage of production
● consider repeatability and use CNC
● explain the quality control measures taken to produce your product
● consider health and safety.

It is important to effectively plan out the production of your prototype and then carry out that plan correctly. As part of this, you should list all of the stages needed to produce your prototype and explain them in detail. You should show how you applied quality control checks to ensure that your product meets the needs of the user.

You should also think about the repeatability of the production process. For example, the use of jigs, fixtures, programming and computer numerical control (CNC) machinery (see Section 6, Practical engineering skills).

Your plan should follow a logical order and cover all health and safety considerations for every stage of production; for example, the personal protective equipment (PPE) that will be needed and the specific safety rules associated with using each piece of equipment.

	Stage/Equipment used/Process	Time Needed	Quality Control	Health and Safety	Photo
Production plan - CNC Milling Machine					
1	● Save your created 3D model of the car boby as an STL file and transfer onto memory stick. ● Plug memory stick into computer that is next to the Milling machine	5 mins	● Make sure your 3D solidworks model is the correct size and scale	● Take regular breaks when on the computer to avoid eye strain.	
2	● Open up the door to the milling machine and make sure the jig is set up with the balsa wood block inside. Push the jig to left until it cant go any further. ● Tigthen the bolts with an allen key. ● Close the door.	5 mins	● Check the balsa wood blood for any imperfections as this good effect the quality of your car. ● Make sure the jig is pushed to the left ● Check the bolts on the jig are tight enough	● Make sure the CNC Machine door is pushed all the way up so that it down fall back down onto you when you are setting up the jig.	
3	● Open up the software on the computer and set up a raster roughing and raster finishing process. ● Start the processes. ● Once the machine has out one side of the car switch it over and repeat the raster process, repeat again for the bottom of the car.	10 mins 1-3 hrs	● Make sure the fininshing amount of the rather processes is the same for each out you do of the car. ● When turning the car over take care not to damage the balsa wood and make sure the jig is pushed to the left.	● Take care when turning the car as there could be some splinters or sharp edges on the balsa wood.	
4	● When the final process is done open the door and remove the jig. ● Remove the car out of the jig.	5 mins	● Take care not to damage the balsa wood when removing it from the machine	● Take care when unscrewing the bolts on the jig and when handling the jig.	
5	● Using sand paper, sand down all rough on the car and make sure the surfaces are smooth	20 mins	● Make sure you sand each side of the car evenly so they are the same.	● When sanding the balsa avoid breathing in the dust that is airborne.	

Figure 7.2.6 **Production plan for the use of a CNC milling machine**

An example production plan for the use of a CNC milling machine is shown above. This contains a detailed description of each stage required to make the product using this piece of equipment. The time needed for each stage is included, along with required quality control checks and health and safety considerations. When presented with your NEA problem, think about the stages that you will need to manufacture the product and how they can be broken down along similar lines. You can also leave space for photographs to provide evidence of how successful each stage was in practice.

How will I be assessed?

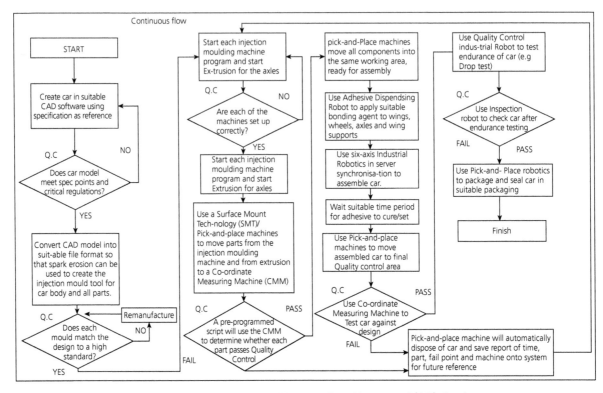

Continuous flow

START

Create car in suitable CAD software using specification as reference

Q.C — Does car model meet spec points and critical regulations? — NO

YES

Convert CAD model into suit-able file format so that spark erosion can be used to create the injection mould tool for car body and all parts.

Q.C — Does each mould match the design to a high standard? — NO — Remanufacture

YES

Start each injection moulding machine program and start Ex-trusion for the axles

Q.C — Are each of the machines set up correctly? — NO

YES

Start each injection moulding machine program and start Extrusion for axles

Use a Surface Mount Tech-nology (SMT)/Pick-and-place machines to move parts from the injection moulding machine and from extrusion to a Co-ordinate Measuring Machine (CMM)

Q.C — A pre-programmed script will use the CMM to determine whether each part passes Quality Control — PASS

FAIL

pick-and-Place machines move all components into the same working area, ready for assembly

Use Adhesive Dispending Robot to apply suitable bonding agent to wings, wheels, axles and wing supports

Use six-axis Industrial Robotics in server synchronisa-tion to assemble car.

Wait suitable time period for adhesive to cure/set

Use Pick-and-place machines to move assembled car to final Quality control area

Q.C — Use Co-ordinate Measuring Machine to Test car against design — PASS

FAIL

Use Quality Control indus-trial Robot to test endurance of car (e.g Drop test)

Q.C — Use Inspection robot to check car after endurance testing — PASS

FAIL

Use Pick-and- Place robotics to package and seal car in suitable packaging

Finish

Pick-and-place machine will automatically dispose of car and save report of time, part, fail point and machine onto system for future reference

Figure 7.2.7 Production flowchart showing the application of quality control (QC) checks

A flowchart is another way of showing how quality control checks could be implemented. The example above uses decision boxes to show the checks needed at each stage of production.

Engineering skills used (15 marks)

For this section, you will need to:
- demonstrate your ability to safely use a range of materials, parts, components, tools and equipment when making your product
- explain your choices of processes used
- explain and provide evidence of your quality control measures and working to tolerances.

This section covers how skilfully you are able to use materials, tools and equipment and tolerances. You should be able to justify the choices that you have made at each stage of your production. For example, if you chose to use a manual milling machine for one part of the product, you should be able to say why it was the most appropriate piece of equipment to use at that stage compared to other potential options, such as a CNC milling machine, manual lathe or hand tools.

You should also explain each process that you used to produce your product and produce evidence of your planned quality control checks being applied. Your finished prototype will need to show high levels of quality throughout. This will involve making sure your product is within acceptable tolerances. More information on quality and tolerances can be found in Section 6 of this book.

Record of production - Car Main Body

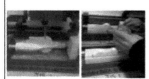

After designing the main body on solid-works we saved is an STL file and transferred it to the computer beside the CNC milling Machine. We then set the jig up, which involved placing it onto a surface plate and using a set square to make sure the balsa block was in straight.

Once the jig was set up, screwed securely into the machine and checked that everything was set we closed the door of the machine. On the computer we then opened the STL file of the car body into the Denford software, we then set the raster roughing and finishing processes.

Once the raster roughing and finishing processes were set we ran them. We also had a vacuum nozzle slotted above the drill piece to clear any balsa wood dust and debris. After one side of the car was cut we rotated it in the jig and ran the same processes. We repeated this until the whole of the car was cut.

After the car was milled out we sanded it down so it was smooth, we then hand painted 3-4 coats of sanding sealer before again sanding it down to a smooth finish. We then repeated this process but sprayed on coats of wood filler primer, after sanding these coats down to a smooth finish it was then ready for its first coat of white gloss paint.

We sprayed the primed car with a white gloss paint, at first we done a few light coats to get an even and smooth finish. Once these had dried we then began a couple of heavier coats. Once we had sprayed these layers we used strip tape to get a clean line, using everyday masking tape would let the paint run,

Once the car was taped up neatly and the only exposed part was the canister we were ready for the coats of black gloss paint. We repeated the same process of how we applied the white gloss to the exposed canister.

Once the paint had dried we carefully removed the tape, taking care not to pull the paint off of the car. We also used a paint brush to clean up any rough edges where paint had faded etc.

The final steps to making the car body were to print off our sponsors logos onto water transfer paper, once we done this we applied them and let the car dry so that the water in the transfers evaporated leaving the decal stuck onto the car body. After the car had been decaled we applied a thin coat of lacquer to begin. Once we knew how much lacquer we needed to apply for the correct weight we did so with care.

Figure 7.2.8 **Record of production for the main body of a racing car**

One way of documenting your production process is a production record (see above). This could describe the steps taken and include photographic evidence. This example describes the tools and equipment used in this particular project including jigs and CNC equipment used.

Applying systems technology (10 marks)

For this section, you will need to:
- demonstrate your understanding of the systems you have used and produce block diagrams to represent them
- explain your use of any technological systems in your product, using diagrams.

This section is about showing your understanding of systems. You should use block diagrams to represent the systems that you have used in your product and thoroughly explain how they work using technical language. For example, if creating an electronic system as part of your product you should use the correct technical names for each component used, such as light-dependent resistor, PIC microcontroller, light-emitting diode, etc.

You should aim to use and explain in detail two or more systems that ensure the product functions as intended. These could be electrical, electronic, mechanical, structural or hydraulic/pneumatic systems. Depending on the brief, you might choose to use a mixture of these different types to achieve your solution. For example, a robot arm used to move objects on a production line would require a programmable system to control the movement of the arm. It could also use pneumatic and/or mechanical systems for a grabber hand that lifts the objects.

Testing and evaluating (10 marks)

For this section, you will need to:
- demonstrate your ability to undertake appropriate testing of your product
- evaluate your product's effectiveness

How will I be assessed?

- undertake an analysis and provide an honest evaluation of your product, making recommendations for improvements.

Finally, you will need to test how well your product works and write an evaluation of how effective it is. For this evaluation, you should consider how well your finished prototype solves the problem set out in the original brief. Your evaluation should also cover how your solution could be improved further if the problem were to be revisited. It should include comments related to both the operation of the systems used and the manufacturing processes followed.

You should provide evidence of detailed and objective (unbiased) testing carried out and the results of the testing. This could be done in the form of a table listing each test, why it was used, how it was carried out, what happened during the testing and what was learnt from it. The tests could include dimensional measurements, functional tests and visual inspections. Assessment of tolerances will play a big role too, as it is important that you check whether the finished product falls within those set earlier in the design process. For example, if the length of a 20 mm part needs to be within 0.5 mm of this value, measurements should be made to ensure that this is the case.

An effective evaluation will build on the results of testing. For example, a visual test might identify surface finish flaws that could be improved upon if the product was made again. A test using measuring equipment might find that one part is slightly too long and out of tolerance. The effectiveness of your evaluation will also depend on how comprehensive your original specification was. You will struggle to conduct useful testing if there are few testable criteria in your specification!

Try to be honest when suggesting improvements. No product is ever completely perfect and there will always be something that you could do better next time. It is also important to note that evaluation is not restricted to the final stages of the project. You should look to evaluate throughout your project so that improvements can be made as you go along. Doing this will help you to achieve a successful end result.

Quality control - Tolerance check

T 3.3 : Length of car	Min : 170mm Max : 210mm	209.50mm	39.50mm
T 3.4 : Width of car	Min : 60mm Max : 85mm	65mm	5mm
T 3.5 : Overall height of car	Max : 60mm	54mm	No error margin
T 3.6 : Total weight of car	Min : 52grams	51.50g	0.50g
T 8.1 : Four Wheels and share common axles	Min : 4 Wheels Max : 4 Wheels	4 wheels	No error margin
T 8.2 : Diameter of front wheel	Min : 26mm Max : 34mm	26.50mm	0.50mm
T 8.2 : Diameter of Rear wheel	Min : 26mm Max : 34mm	26.50mm	0.50mm
T 8.3 : Width of front wheel	Min : 15mm Max : 19mm	15mm	No error margin
T 8.3 : Width of rear wheel	Min : 15mm Max : 19mm	15mm	No error margin
T 10.9.1 : Front wing span	Min : 40mm Max : 85mm	40mm	No error margin
T 10.9.2 : Rear wing span	Min : 40mm Max : 65mm	40mm	No error margin

As you can see above we scored 96/120 in the specification judging because we broke a critical regulation, but tolerance wise we were fine, everything was within tolerance of the specification.

The reason why a lot of the tolerances are "No error Margin" is because we tried to design everything to the minimal or maximum it could be depending on how it benefitted the car. For example we made the width of the wheels exactly 15mm because that means there will be less surface friction. We also made the diameter as close to 26mm as possible taking into consideration the excess material that could be taken off by the lathe.

Figure 7.2.9 Checks to determine whether tolerances have been met

The figure above shows an example of a list of tolerances that needed to be checked for a prototype racing car. An explanation has been given as to why the tolerances were selected and the outcome of testing against them. Think about the tolerances that will be needed for the different parts of your project. How will you ensure that they are met?

ENGINEERING EQUATIONS AND SYMBOLS

Below is a list of the equations you will be expected to be able to recall and use within the exam.

Description	Calculation	Equation
Area of a rectangle	Area = length × width	$A = L \times W$
Area of a circle	Area = pi × (radius × radius)	$A_c = \pi r^2$
Area of a triangle	Area = half (base × height)	$A_t = \frac{1}{2}(B \times H)$
Volume of a cuboid	Volume = length × width × height	$V = L \times W \times H$
Volume of a cylinder	Volume = area of circle × length	$V_c = A_c \times L$
Density	Density = mass / volume	$\rho = m / V$
Stress	Stress = force / cross sectional area	$\sigma = F / A$
Strain	Strain = change in length / length	Strain $= \delta l / l$
Young's modulus	Young's modulus = stress / strain	$E = \sigma / \varepsilon$
Pressure	Pressure = force / area = F / A	$P = F / A$
Factor of safety	Factor of safety = yield stress / load	$FoS = \sigma_y / L$
Ohms law	Current = voltage / resistance	$I = V / R$
Resistance of resistors in series	Total resistance = sum of all resistances in series	$R_T = R_1 + R_2$
Resistance of resistors in parallel	1/ total resistance = sum of (1 / each resistance in parallel)	$1 / R_T = 1 / R_1 + 1 / R_2$
Gear ratio	Gear ratio = number of teeth on driven gear / number of teeth on driver gear	Gear ratio $= N_{driven} / N_{driver}$
Velocity ratio	Velocity ratio = size of output wheel / size of input wheel	Velocity ratio $= d_{output} / d_{input}$
Mechanical advantage	Mechanical advantage = load / effort = F_b / F_a	$MA = F_b / F_a$

Symbol	Description
=	Equal to
<	Smaller than
≤	Smaller than or equal to
>	Greater than
≥	Greater than or equal to
±	Plus–minus
∝	Proportional to
≈	Almost equal to
π	Pi
ρ	Density
σ	Stress
ε	Strain

INDEX